Cape gooseberry

The Authors

Dr. Desh Beer Singh is presently Principal Scientist (Horticulture) and Head VC &F at ICAR-Central Institute of Temperate Horticulture, Srinagar; J&K. He did his M.Sc (Horticulture) from Punjab Agricultural University, Ludhiana, Punjab and Ph.D (Horticulture) from Bhimrao Ambedkar University, Agra. His field of specialization is production and post harvest management of horticultural produce and its value addition. Further he has obtained advanced training from USDA, USA in the field of conservation of plant genetic resources and from RAS, Moscow (Russia) in the field of post harvest management of horticultural produce. Dr. Singh has also identified a new species of banana and several unexploited tribal fruits from forests of Andaman and Nicobar Islands.

He has published more than 110 research papers and contributed 14 bulletins, 15 book chapters, 05 books besides 55 presentations in national and international conferences (2 presentations awarded as best paper). Dr Singh has also guided students for their master's degree in Horticulture, Food Technology and M. Phil programme. He has been a part time faculty member in Institute of Business Management, Rajasthan Agricultural University, Bikaner, Rajasthan and was also an invited member in Board of Post Graduate Studies, Nagaland University, Nagaland. He is also a life member of 10 professional societies where he is associated as editor, executive member, councillor and referee of journals and societies. Dr. Singh commercialized the pineapple cultivation in Andaman and Nicobar Islands and has been rewarded by SAARC Agricultural Information Centre for his success story on production of pineapple round the year in Andaman and Nicobar Islands. Besides associated for development and release of 04 varieties of temperate fruits, Dr. Singh has developed and standardized 06 processing technologies for making novel value added products of temperate fruits out of these three technologies are licensed and commercialized by entrepreneurs. Dr. Singh is recipient of Team Award from ISAE for development of Pomegranate Aril Extractor, Fellow of Horticultural Society of India, Fellow Hind Horticultural Society, and Bioved Fellowship from Bioved Research Society, Allahabad.

Prof. Nazeer Ahmed is serving as Director at ICAR-Central Institute of Temperate Horticulture (CITH), Srinagar and before that he was at SKUAST (K), Srinagar where he worked for 22 years at various capacities as Scientist, Senior Scientist, Associate Director Research, Professor and Head, Dean Agriculture and also as Director Resident Instruction cum Dean Post Graduate Studies. During 29 years of service, he has contributed significantly as a teacher, researcher, extension worker and administrator. As teacher, he taught and designed both UG and PG courses and guided more than hundred Master's and PhD's as Chairman and Co-Chairman and was instrumental in establishment of Faculty of Horticulture at SKUAST (K). Through various ad-hoc projects he created state of art field and laboratory facilities both at the university and at CITH and published more than 300 research papers, extension bulletins and review papers in national and international journals. During the period, he attended more than 90 workshops and conferences and conducted several national trainings, summer/winter schools and workshops as Course Director. In research, he executed several multidisciplinary and inter-institutional, national and state projects and developed and released 33 varieties and hybrids and recommended 64 production recommendations for temperate fruits and vegetables, which have covered significant area inn temperate region contributing to national income.

As a horticulture expert, he executed number of special assignments as Chairman, Co-Chairman and Member of various committees, international delegation and conferences. He is a recipient of Dr. R.S. Paroda Award; Dr. M.H. Marigowda National Endowment Award; Dr. Kirti Singh Gold Medal; Vijay Shree Award; Glory of India Gold Medal and Chaudhary Devilal Outstanding award for best AICRP (VC & Potato) for his contribution to agricultural research and education in temperate region and is a fellow of Indian Society of Vegetable Science, Horticultural Society of Indian Society of Horticultural Research and Development and confederation of Horticultural Association of India besides having several best paper awards. He is also a life member of sixteen professional societies where he is associated as Patron, Chief Editor, Executive Councillor and referee of journals and societies.

Cape gooseberry

Authors

Desh Beer Singh

Nazeer Ahmed

2015

Daya Publishing House®

A Division of

Astral International Pvt. Ltd.

New Delhi – 110 002

Cataloging in Publication Data--DK
  Courtesy: D.K. Agencies (P) Ltd. <docinfo@dkagencies.com>

  Desh Beer Singh, author.
  Cape gooseberry / authors, Desh Beer Singh, Nazeer Ahmed.
    pages   cm
    Includes bibliographical references (pages   ) and index.
    ISBN 978-93-5130-678-8 (International Edition)

    1. Cape gooseberry--India. 2. Cape gooseberry.   I. Ahmed, Nazeer
    (Horticulturist), author. II. Title.

  DDC 634.230954   23

Published by   : **Daya Publishing House®**
                        A Division of
                        Astral International Pvt. Ltd.
                        – ISO 9001:2008 Certified Company –
                        4760-61/23, Ansari Road, Darya Ganj
                        New Delhi-110 002
                        Ph. 011-43549197, 23278134
                        E-mail: info@astralint.com
                        Website: www.astralint.com

Laser Typesetting   : **Classic Computer Services**, Delhi - 110 035

Printed at   : **Thomson Press India Limited**

PRINTED IN INDIA

Preface

Systematic introduction and adaptation contributes increase in the diversity of agricultural systems and provides new alternatives to producers and processers. The increasing demand for their diversified crops, together with environmental concerns particularly that of climate change, makes evaluation and standization of sustainable agriculture production techniques a necessity.

Among these alternative crops, Cape gooseberry, grown for edible fruits are rich in vitamin C, A, B and polyphenols which can be eaten raw, processed for various value added products has wide adaptation and can be successfully grown in temperate regions during summers.

Lack of scientific production management techniques and inadequate processing technology for making novel value added products are the main reasons for its less expansion in temperate region.

This book is an attempt to put in one place a synopsis of worldwide research findings and information and possible crop production practices. This book integrates production and processing of Cape gooseberry covering all aspects in simple and comprehensive manner

It is hoped that information produced in the book will be quite useful to farmers, processors, students, researchers and all concerned with environment and alternate crop use, production and processing of Cape gooseberry.

Help, support and suggestion given by my colleagues and field workers of Central Institute of Temperate Horticulture, Srinagar are duly acknowledged.

Desh Beer Singh

Nazeer Ahmed

Dr. N. K. Krishna Kumar
Deputy Director General
(Horticultural Science)

भारतीय कृषि अनुसंधान परिषद
कृषि अनुसंधान भवन–II
पूसा, नई दिल्ली–110 012

Indian Council of Agricultural Research
Krishi Anusandhan Bhawan-II
Pusa, New Dellhi-110 012

Foreword

Growing consumer demand for new and nutrient rich fruits and vegetables are spurring a need for increased information on production technologies of various crops. The increasing demand for these new and alternate crops, together with environmental concerns particularly that of climate change, makes standardization and evaluation of sustainable agricultural production techniques a necessity.

Systematic approaches on introduction and adaptation of crops and varieties contributes to increase in the diversity of agricultural systems and offers alternatives to stakeholders with crops that may have better value and further add to their knowledge. There are many crops of tropical and sub-tropical regions that have desirable attributes to be introduced as alternate crops in temperate climates. Cape gooseberry is annual crop and its fruits are rich in vitamins A, B, C and polyphenols. Its fruits can be eaten raw, as dessert and appetizer and can also be utilized for preparing jam and osmo dehydrated slices. Due to wide adaptation and varied uses of fruits it has good prospects for expansion as a new cash crop in temperate regions.

The book *Cape gooseberry* is a kind in itself and an attempt to put enormous scientific information on its production and utilization in single compilation like this one for the benefit of end users. I appreciate the initiatives taken by the authors of the book and wish that it serves its purpose.

(N. K. Krishna Kumar)

Contents

Chapter 1

Introduction and History

Cape gooseberry (*Physalis peruviana*), known in English as a golden berry (South Africa). Inca berry, cape gooseberry, giant ground cherry, Peruvian cherry (U.S.), poha (Hawaii), ras bhari (India), aguaymanto (Peru), uvilla (Ecuador), uchuva (Colombia) and *Physalis.* It is indigenous to South America but was cultivated in south Africa in the region of the Cape of Good Hope during the 19th century, imparting the common name, "Cape gooseberry". Cape gooseberries are well known for its blood purifying capacity. They are also known for other medicinal qualities which is being a source of provitamin A, vitamins B and C. In India scientists have isolated physalolactone C from the leaves, a minor steroidal constituent removed from the paper-like husks. The attractive yellow marble-sized fruit makes an extremely tasty jam. Fruit has a high vitamins A, B, and C content, is a rich source of carotene, phosphorus and iron, and also contains vitamin P. It may be eaten fresh, in salads or in cocktails.

Ethnic names - Jam fruit, Peruvian cherry, Uchuva. Scientific name *Physalis peruviana* was derived from the Greek word physa meaning bladder, for the calyx covering the fruit and peruviana meaning of Peru.

In the 18th Century, the fruits were perfumed and worn for adornment by native women in Peru. They are also widely used in folk medicine, and grow wild in Europe, America and Asia. *Physalis peruviana,* a South American species, is being used as a natural medicine in the tropical countries where it grows since a long time ago. Its use by rainforest Indians in the Amazon is well documented, and its edible sweet-tart fruits are enjoyed by many rainforest inhabitants, animal and human alike.

Four species are cultivated in different parts of the world for their fruit: *Physalis peruviana* (Cape gooseberry, uchuba) and *Physalis pruinosa* (ground cherry, husk tomato) are used as jam fruits; *Physalis alkekengi* L. (Chinese lantern) is used as an ornamental; and *Physalis ixocarpa* (tomatillo, tomate de cascara) is used as a vegetable or for sauces. *Physalis peruviana* has been cultivated since pre-Columbian times along the Andes Mountains of South America. Their fruits are smaller and sweeter than *tomatillos* and can be eaten raw or used in preserves. It has a particularly delicious fruit with a tangy pineapple-like flavor. These plants grow all over the Andes and were fruit of the Incas (Veitmeyer, 1991). The fruit is eaten fresh or cooked. They make excellent pies and jellies and are very high in pectin. The fresh fruit may be served with husk pulled back for fondue. *Physalis peruviana* sauce is a nice accompaniment to a meat dish. While not well known by the retail consumer, the fruit has a strong following among chefs and the market is likely to grow for good quality, vine-ripened fruit.

This plant species has been spread by explorers and travelers worldwide, but is still considered a backyard fruit in most areas. Small industries are developing around *Physalis peruviana* countries in Central and South Africa, Australia, New Zealand, and India but nowhere has it really achieved large commercial success (Morton, 1987). The plant's productivity in poor soils, its ease of cultivation, and low requirement for water and fertilizer has made it an attractive potential crop. *Physalis peruviana* was grown by early settlers at the Cape of Good Hope before 187. In South Africa it is commercially cultivated and common as an escape and the jam and canned whole fruits are staple commodities, often exported. It is also cultivated and naturalized on a small scale in Gabon and other parts of Central Africa. Soon after its adoption in the Cape of Good Hope it was carried to Australia and there acquired its common English name. It was one of the few fresh fruits of the early settlers in New South Wales. There it has long been grown on a large scale and is abundantly naturalized, as it is also in Queensland, Victoria, South Australia, Western Australia and Northern Tasmania. It was welcomed in New Zealand where it is said that "the housewife is sometimes embarrassed by the quantity of berries [Cape gooseberry, *Physalis peruviana*] in the garden," and government agencies actively promote increased culinary use.

In China, India and Malaya, *Physalis peruviana* is commonly grown but on a lesser scale. In India, it is often interplant with vegetables. It is naturalized on the island of Luzon in the Philippines.

Seeds were taken to Hawaii before 1825 and the plant was naturalized on all the islands at medium and somewhat higher elevations. It was at one time extensively cultivated in Hawaii. By 1966, commercial culture nearly disappeared and processors has to buy the fruit from backyard growers at high prices. It is widespread as an exotic weed in the South Sea Islands but not seriously cultivated. The first seeds were planted in Israel in 1933. The plants grew and bore very well in cultivation and soon spread as escapes, but the fruit did not appeal to consumers, either fresh or preserved and promotional efforts ceased. In England, *Physalis peruviana* was first reported in 1774. Since that time, it has been grown there in a small way in home gardens and after World War II was canned commercially to a limited extent. Despite this background, early in 1952, the Stanford Nursery, of Sussex, announced the "Cape gooseberry, the wonderful new fruit, especially developed in Britain by Richard I. Cahn". Concurrently, jars of *Physalis peruviana* jam from England appeared in South Florida markets and the product was found to be attractive and delicious. It is surprising that this useful little fruit has received so little attention in the United States in view of its having been reported on with enthusiasm by the late Dr. David Fairchild in his well-loved book, "*The World War by Garden*". He there tells of its fruiting "enormously" in the garden of his home, "In The Woods", in Maryland, and of the cook's putting up over a hundred jars of what he called "Inca Conserve" which "meet with universal favor". It is also remarkable that it is so like known in the Caribbean islands, though naturalized plants were growing profusely along roadsides in the Blue Mountains of Jamaica before 1913. With a view to encouraging *Physalis peruviana* culture in Florida, the Bahamas, and the West Indies, seeds have been repeatedly purchased from the Stanford Nursery and distributed for trial. Good crops have been obtained. Nevertheless there was no incentive to make further plantings.

Physalis pruinosa, a closely related species, is the most popularly grown variety because at 45 cm to 50 cm (18 to 20 inches) high and 60 cm to 90 cm (2 to 3 feet) wide, it is smaller and more manageable than *Physalis peruviana*. The fruits are a lot like those of *Physalis peruviana* in flavor and size, but the husks are tight fitting and they curl back to expose the ripe fruits. The small yellow fruits of *Physalis pruinosa* are used for sauces, pies and preserves in mild-temperate climates. However, *Physalis peruviana* is said to bear a superior fruit and has become widely known.

Physalis ixocarpa, a closely related species native to Mexico, was a prominent staple in Aztec and Mayan economy. The plant abounds in Mexico and the highlands of Guatemala and the fruits are commonly seen in native markets. In Mexico and other countries in Latin America, the fruits of *Physalis ixocarpa* are crushed and used in sauces. They have an agreeable, but not sweet flavor. *Physalis ixocarpa*, is gaining ground as a new crop in California due to increased popularity of Mexican food in the United States (Quiros, 1984) and has production potential in the Southern United States. In Louisiana, *tomatillo* imported from Mexico is sold as a fresh fruit in few grocery stores. There is a potential market for fresh produce and the Louisiana sauce industry may be interested in opening a new ethnic market for their products. Nevertheless, this species has not been as widely distributed abroad as *Physalis peruviana*. It was introduced into India in the 1950's and is cultivated in the northwest desert region of Rajasthan. In Queensland, Australia and in South Africa it has fruited prolifically. There is some commercial cultivation in Pietersburg, South Africa, for processing. It was too-successfully introduced into East Africa, for, in 1967, it was reported to be the important weed of agricultural fields in the highlands of Kenya.

Physalis philadelphica Lam., a closely related species, is cultivated in Mexico and Guatemala and originating from Mesoamerica. Various archaeological findings show that its use in the diet of the Mexican population dates back to pre-Columbian times. Indeed, vestiges of *Physalis* sp. used as food have been found in excavations in the valley of Tehuacan (900 BC-AD 1540). In pre-Hispanic times in Mexico, it was preferred far more than the tomato (*Lycopersicon* sp.). However, this preference has not been maintained, except in the rural environment where, in addition to the persistence of old eating habits, the tomato's greater resistance to rot is still valued. Possibly because of the fruit's colorful appearance and because there are ways of eating it which are independent of the chili (*Capsicum* sp.), the tomato achieved greater acceptance outside Mesomerica and *Physalis* sp. was marginalized, or its cultivation was discontinued, as happened in Spain.

Physalis philadelphica was domesticated in Mexico from where it was taken to Europe and other parts of the world; its introduction into Spain has been well documented. Indeed, it is believed that this species originated in central Mexico where, at present, both wild and domesticated populations may be found. The fruit of *Physalis chenopodifolia*, other closely related species, is picked in the state of Tlaxcala Mexico. In Europe, *Physalis alkekengi* is grown as an ornamental plant because of the colorful calyx of its fruit, and its fruit also is used in central southern Europe. *Physalis chenopodifolia* Lam. is in the initial stage of domestication and shows a favorable response to agricultural practices; accordingly, it must be collected and evaluated so that the potential for better utilization in the future may be established.

Physalis ixocarpa and *Physalis philadelphica* have been a constant component of the Mexican and Guatemalan diet up to the present day, chiefly in the form of sauces prepared with its fruit and ground chilies to improve the flavor of meals and stimulate the appetite. They are also used in sauces with green chili, mainly to lessen its hot flavor. The fruit of *Physalis ixocarpa* is used cooked, or even raw, to prepare purees or minced meat dishes which are used as a base for chili sauces known generically as *salsa Verde* (green sauce); they can be used to accompany prepared dishes or else be used as ingredients in various stews. An infusion of the husks (calyx) is added to tamale dough to improve is spongy consistency, as well as to that of fritters: it is also used to impart flavor to white rice and to tenderize red meats. In Mexico, fruits are used in the making if chili sauce and dressing for popular such as tacos and enchiladas. About ten years ago the crop began to be industrialized in Mexico and agro-industries are currently estimated to process 600 tons per year, 80 per cent of which is exported to the United States as whole *tomatillos*, without a calyx and canned, while the remainder is used in the preparation of packaged sauces for the domestic market. *Physalis philadelphica* is acquiring importance as an introduced crop in California as a result of the growing popularity of

Mexican food in the United States. Furthermore, numerous medicinal properties are attributed to it. Official statistics show that, in 1984, 1548 ha were in Mexico, with a total production value of 5797 million pesos and an average per caput consumption of 2.32 kg. Both in Mexico and Guatemala, wild tomato fruit from cultivated fields has a predominant place in the diet. Hence in some regions it is an important product among those gathered in rural areas for immediate consumption and for sale.

Chapter 2
Geographic and Cultivation Origins

Many annual herbs within the genus *Physalis* are indigenous to many parts of the tropics, including the Amazon. *Physalis peruviana* L., a South American species, is native to the Amazon basin (Brazil and Peru), although it has spread and naturalized to highland areas of Peru and Chile. *Physalis minima* L., a closely related species, native to India and other parts of tropical Asia. *Physalis ixocarpa* Brot. ex Hornem., other closely related species, is a native of Mexico. *Physalis angulata* L. and *Physalis pubescens* L., two other closely related species, are native to Tropical America. Native to high altitude tropical Colombia, Chile, Ecuador and Peru where the fruits grow wild, physalis are casually eaten and occasionally sold in markets. It is found in markets from Venezuela and Chile has long been a minor fruit of Andes (FAO 1982).Only recently has the plant become an important crop; it has been widely introduced into cultivation in other tropical, subtropical and even temperate areas. Legge (1974) mentioned that it is native to Peru, in the same areas where tomato originated where as, Bartholomaus *et al.* (1990) reported that it comes from Ecuador and Peru. There are indications that fruit came from Braziland was acclaimited in Peru and Chile (CRFG, 1997).

Soon after its adoption in the Cape of Good Hope (presumably the origin of the name 'Cape gooseberry'), it was carried to Australia, where it was one of the few fresh fruits of the early settlers in New South Wales (Morton 1987. It is also favored in New Zealand where it is said that "the housewife is sometimes embarrassed by the quality of berries in the garden", and government agencies promote increased culinary use. It is also grown in India, and it is called Rasbhari in Hindi. The Cape gooseberry is also grown in North Eastern China, namely Heilongjiang province. A seasonal fruit harvested in late August through September. In Chinese pinyin, the fruit is informally referred to as "gu niao" and the scientific name is _Physalis pubescens_ L. or in Chinese pinyin "mao suan jinag". It has been widely grown in Egypt for at least half a century and is known locally as "harankash" a word of obscure origin, or as is-sitt il- mista Hiya (the shy women), a reference to the papery sheath. It makes an excellent crumble, substituting harankash for apples.

Wild Cape gooseberry is commonly found on the bounds of the fields, wastelands, around the houses, on roadsides, etc., where the soil is porous and rich in organic matter. It is annual herbaceous plant having a very delicate stem and leaves. It is found growing in the sub-Himalayas up to attitudes of 1,650 meters. It is also reported to grow in Afghanistan, Baluchistan, tropical Africa, Australia. Ceylon, etc. A small, delicate, erect, annual, pubescent herb, 1.5 meters tall; intermodal length, 8.2 cm; more or less the whole plant is pubescent. Leaves, petiolate (4.1cm long), ovate to cordate, pubescent, delicate, exstipulate, acuminate, having reticulate palmate venation and undulate margins; dorsal surface of the leaves, dark green and the ventral surface, light green; 9.7 cm long and 8.1 cm broad. Flowers, pedicellate having 1.2 cm long pedicel, hermaphrodite, complete solitary small companulate, 1.2 to 1.4 cm in diameter; calyx; gamosepaleous, 5-toothed, actinomorphic, green, persistent, downy; corolla, gamopetalous with five petals, the petal cup, 1.1 to 1.3 cm long, yellow, having five black stops on yellow ground in the middle of the corolla cup; stamens, five epipetalous, 6 to 7 mm long, having a black filament and greenish yellow anther lobes; style, black, 9 mm long, having a yellowish stigma at the top and a yellowish round ovary at the base. Fruit, a berry, enclosed within the enlarged, 10-ribbed, reticulately veined calyx, which is 4.1 cm long and 2.5 cm broad; berries, stalked (stalk, 2.2cm long), almost round having a pinhead-sized depression at the end; diameter, 1.4 to 1.6 cm; weight, 2.15g; volume, 1.32 ml; fully mature fruits primrose yellow 601/2 at full maturity. Seeds, globose, Dresden yellow 64/3; weight and volume of 100 seeds, 113 mg and 197 micro-liters respectively.

Chapter 3

Distribution

Physalis peruviana can be found in most continents in the tropics, including Africa, Asia and the America. It has naturalized in tropical regions around the world (including Hawaii) and it has been widely introduced into cultivation in other tropical, subtropical and even temperate areas. It is said to succeed wherever tomatoes can be grown. In South America, *Physalis peruviana* has been reported in Peru, Brazil, Chile, Ecuador, Colombia and Venezuela. In New Zealand *Physalis peruviana* plants are grown on a few small scale. Production is small and fruit is supplied mainly to the local market. This plant species is also present on Pacific Islands. According to Duthie (1905), *Physalis minima,* a closely related species native to India, also grows in Afghanistan, Baluchistan, tropical Africa, Australia, Ceylon, etc. In the United States, *Physalis peruviana* is hardly seen, except in Pennsylvania (Dutch county and parts of the Midwest). In Haiti, the presence of *Physalis angulata* and *Physalis pubescens* L. has been reported. The *Physalis philadelphica* plant, other closely related species, grows from Southern Baja California to Guatemala, from 10 m in Tree Valles, Veracruz, to 2600 m in the valley of Mexico.In several regions of Mexico, the species *Physalis chenopodifolia* Lam. grows wild in cultivated fields: its use as a potential resource has been recorded.

Several species of *Physalis* are widespread in America as endemic weed species. Six important *Physalis* spp. are prevalent in the phytogeographic region of Mesoamerica (Belize, Guatemala, Honduras, El Salvador, Nicaragua, Costa Rica and Panama and the Mexican states of Chiapas, Yucatan and Quintana Roo): *Physalis angulata, Physalis cordata* Mill., *Physalis gracilis* Miers, *physalis ignota* Britt, *Physalis lagascae* R. and S., and *Physalis pubescens* L. (Gentry and D'Arcy 1986). These *Physalis* spp. can be intercrossed, but incompatibility has been found (Pandey 1957, Quiros 1984). The basic chromosome number of the genus is N=12 and most species are diploid. *Physalis peruviana* is a tetraploid (Menzel, 1951). Wild Cape gooseberry is commonly found in the bunds of the field, water land and the house, and on the road sides where the rain is porous and rich in organic matter.

Chapter 4

Botanical Description

4.1 Habit

Physalis peruviana spontaneous clumps of plants can be found along river banks and just about anywhere the soil is distributed and the canopy is broken (allowing enough sunlight to promote its rapid growth). In Hawaii, *Physalis* is naturalized in distributed sites in mesic to wet forest, diverse mesic forest and subalpine woodland, 450-2,020 m (Wagner *et al.,* 1999). In Fiji, *Physalis peruviana* is seen at elevations from near sea level to 900m in gardens and also naturalized in forest along trails and streams, in clearing and cultivated areas (Smith, 1991). In Niue, *Physalis peruviana* is a common weed in some plantations (Sykes, 1970). Uncommon in plantations and waste places but sometimes collected for food (Whistler, 1988). In Tahiti, *Physalis peruviana* is frequent in all the cool valleys to 800 m (Welsh, 1998). *Physalis minima* is commonly found on the bunds of the fields, wastelands, around the houses on roadsides, etc., where the soil is porous and rich in organic matter. It is an annual herbaceous plant having a very delicate stem and leaves. It is found growing in the sub-Himalayas up to altitudes of 1,650 meters. According to Duthie (1905), it also grows in Afghanistan, Baluchistan, tropical Africa, Australia, Ceylon, etc. This is an ornamental plant whose calyx gets enlarge into a kind of membranous cages and after flowering it is being surrounded by a large berry. It has a fairly sweet taste and can be eaten raw or cooked. It is very pleasing to the eye when the five petals of the calyx are opened around the fruit. They have a slight sweet-sour taste. Plants of cape gooseberry grows wild in many parts of Andes *e.g.* Columbian forests 2,200 m above sea level (a s l) (Fischer, 1995) and many parts of the Hawaiian islands at medium and somewhat higher elevations (Morton, 1987), where it is known as "Poha" (Carman, 180-81). This perennial shrubby herb is a soft wooded, somewhat viving plant, normally grows to height of about 1-1.5 m with sympodial branch ramification. The vegetative growing basal stem somewhat woody at its base, after forming 8-12 m, branches twice dichasially so that four generative stems (or leaders are formed (Fischer 2000). If staked or pruned plant height might reach 2 m. Cape gooseberry plants have an indeterminate growth habit, *i.e.* development of new shoots, leaves flowers and fruits are simultaneous. Gooseberries are bitter, even though it is used in sweet preparations. These piquant fruit turns out from a pale green to an amber or gold color as they get ripen and offer a flavor similar to a gooseberry or a tart green grape. Cape gooseberry supplies are imported from Colombia and South America, the places where they grow wild and are gathered. The fruit matches well with meat and savory foods. Under good conditions it may reach 6 ft. The purplish, spreading branches are covered with fine hairs. As the fruits ripen, they begin to drop to the ground, but will continue to mature. Sometimes the stored fruit can be adversely affected by *Penicillium* and *Botrytis* molds. Perennial herb or soft-wooded small herbaceous shrub, erect up to 1 m in length (but occasionally may attain 1.8 m or 6 ft), straggly with age similar to the common tomato. This herb spreads 1 m to 1.2 m (3 to 4 feet) wide. Branches spreading, ribbed, alternate, often purplish. All parts densely pubescent, simple or glandular hairs up to 1 mm long.

Classification

Kingdom	:	Plant
Subkingdom	:	Tracheobionta (Vascular Plants)
Division	:	Magnoliophyta (Flowering Plants)
Class	:	Magnoliopsida (Dicotyledons)
Subclass	:	Asteridae
Order	:	Solanales

Family   :   Solanaceae

Genus   :   *Physalis* L.

Species   :   *Physalis peruviana*

Synonyms

☆ *Alkekengi pubescens* Moench

☆ *Boberella peruviana* (L.) E.H.L. Krause

☆ *Boberella pubescens* (L.) E.H.L. Krause in Sturm

☆ *Physalis chenopodifolia* Lam.

☆ *Physalis edulis* Sims

☆ *Physalis esculenta* Salisb.

☆ *Physalis latifolia* Lam.

☆ *Physalis peruviana var. latifolia* (Lam.) Dunal

☆ *Physalis tomentosa* Medik

Table 4.1: Common Names in different Countries/Regions

Country	Common Names
Spanish: Peru	"aguaymanto", "capuli", "ahuaimanto", "pasa capuli", "tomate Silvestre", "aguayllumantu", "unchuba", "unchuva", "mullaca", "bolsa mullaca", "camapu", "capuli Cimarron", "alquequenjee"
Boliva	"capuli", "motojobobo embolsado"
Chile	"capuli", "tomato", "amor en bolsa", "bolsa de amor"
Colombia	"uvilla", "uchuva", "vejigon", "guchavo"
Ecuador	"uvilla"
Guatemala	"miltomate"
Mexico	"cereza del Peru", "tomate de cascara", "tomate de fresadilla", "tomate milpero", "tomate verde", "tomatillo", "miltomate"
Venezuela	"topotopo", "chechuva"
French	"capuli", "coqueret du Perou", "groseiller du cap"; *Gabon* "alkekengi", "coqueret".
Cook Islands	"tupera"
Creole	"kokmol", "batter autor", "manman lanman".
Fiji	"botebote yadra", "mbotembote yandra", "kospeli", "maulaqua", "maulanggua", "tukiyadra", "tukiyandra"
French Polynesia	"tupere"

Contd...

Table 4.1–*Contd...*

Country	Common Names
Hawaiin	"poha", "pohâ", pa'ina".
Hindi (India)	"taparee", "tiparee", "makowi", (*Physalis minima*, a closely related species): *Andhra Pradesh* "kupanti", "budda", *Bengal* "budamma"; "ban tipariya"; *Gujrat* "parpoti", "popti"; *Himachal Pradesh* "rasbhary"; *Hindi* "tulati pati", *Karnataka* "gudde hannu"; *Kerala* "njodi njotta"; *Maharastra* "chirboti", "dhan mori"; *Tamilnadu* "tholtakalli"
Kiribati	"te baraki", "te bin"
Niue	"manini"
Philippine	"loboloboha"
Tonga	"ku'usi"
Others	"juá-de-capote", "batter-autour", "k'u chih", "urmoa batoto bita", "cecendet", 'dumadu harachan", "nvovo", "polopa", "polopa", "saca-buche", "thongtheng", "tino-tino", "topatop", "wapotok", "poha", "topo topo".

Physalis minima, a closely related species native to India, is a small, delicate, erect, annual, pubescent herb, 1.5 meters tall. Intermodal length is 8.2 cm. More or less, the whole plant is pubescent. *Physalis pruinosa,* a closely related species, is 45 cm to 50 cm (18 to 20 inches) high and 60 cm (2 to 3 feet) wide. It is similar and more manageable than *Physalis peruviana* L. *Physalis ixocarpa,* a closely related species native to Mexico, grows up to 90 cm (3 feet) high and its branches are covered in heart-shaped, elongated leaves with long stems. *Physalis philadelphica,* other closely related species native to Mexico, is an annual of 15 cm to 60 cm; it is subglabrous, sometimes with sparse hairs on the stem. The genus *Physalis* is a group of annual and perennial herbs bearing globular fruits, each enclosed in a bladderlike husk or calyx which becomes papery on maturity. Species within the genus *Physalis* are grown for their fruits and for decoration. Of the more than 70 species (perhaps 100), only a very few of them are of economic value. Among them are the cape gooseberries (*Physalis peruviana*), husk tomato (*Physalis ixocarpa*) Brot. ex Hornem.

4.2 Leaves

Simple, alternate sometimes seemingly opposite, 6 cm to 15 cm (2 $^3/_8$ in to 6 in) long and 4 cm to 10 cm (1 ½ to 4 in) wide, velvety (pubescent), (Everett, 1981) ovate, acuminate or heart-shaped, base cordate, and apex acuminate (pointed). They appear irregularly along the stems (Moriconi *et al.,* 1990), the margins are entire or rarely with a few blunt lobes and randomly-toothed. The petioles are 2 cm to 3 cm long. Leaves of *Physalis minima* are 9.7 cm long and 8.1 cm broad, ovate to cordate, pubescent, delicate, exstipulate, and acuminate, with reticulate palmate venation and undulate, margins. The dorsal surface of the leaves is dark green and the ventral surface, light green. They are petiolate (the petiole is 4.1 cm long). The leaf of lamina of *Physalis philadelphica* is 9 mm to 13 mm x 6 mm to 10 mm. Its apices are acute to slightly acuminate, with irregularly dentate margins and two to six teeth on each side of the main tooth, of 3 mm to 8 mm.

4.3 Flowers

Solitary in the leaf axils, small, perfect, actinomorphic, pentamerous, bell-shaped, cream or yellow, pedicellate, nodding, 2 cm (3/4 in) wide, with 5 dark purple-brown spots in the throat; calyx purplish-green, hairy, 5-pointed 5-lobed, veins often prominent, lobes acuminate-triangular, near 1 cm long, distinct at apex.

The flowers are bell shapped, nodding hermaphrodite (with both male and female organs) and are pollinated by bees and wind. After the flower falls, the calyx expands, ultimately forming a straw-colored husk much larger than the fruit it encloses. The corolla is yellow with well-defined purplish brown spots at base, (FAO

1982) 15 mm to 20 mm in diameter, the limb rotate or shallowly 5-lobed, the tube swollen into shallow nectar pouches between the filaments, densely pubescent with pale yellowish dendritic hairs below the spots and around the nectarines. The style is 5 mm to 7 mm long. The flowers of *Physalis minima* are pedicellate having 1.2 cm long pedicels. They are hermaphorodite, complete, solitary, small campanulate, 1.2 cm to 1.4 cm in diameter. The calyx or husk which is small at the beginning of fruit development grows to a bladder like organ, which completely encloses the ripening fruit. The corolla is campanulate and gamopetalous with five petals cup, 1.1 cm to 1.3 cm long, yellow, having five black or purple to purplish brown to spots on yellow ground in the middle of the corolla cup. The stamens are five, epipetalous, 6 mm to 7mm long, having a black filament and greenish-yellow anther lobes. The style is black, 9 mm long, having a yellowish stigma at the top and a yellowish round ovary at the base. The flowers appear in acropetal succession, *i.e.* the lower flowers appear and from fruits earlier than the upper ones, which emerges as well as set fruit later. The corolla of *Physalis philadelphica* is 8 mm to 32 mm in diameter, yellow and sometimes has faint greenish blue or purple spots. The anthers are blue or greenish blue. The calyx is accrescent, reaching 18 mm to 53 mm x 11 mm to 60 mm in the fruit, and has ten ribs. The fruit is 12 mm to 60 mm x 10 mm to 48 mm in size and sometimes tears the calyx.

4.5 Fruit

Cape gooseberry fruit is a berry which looks like a marble with a small round shape. The fruit consists of numerous small yellow seeds. The fruit looks orange or yellowish in color and becomes bright yellow and sweet when ripe. The fruit is covered by wrinkled leaves, which form a Chinese lantern. Cape gooseberries are well known for its *blood purifying* capacity. They are also known for other medicinal qualities which is being a source of provitamin A, vitamins B and C. In India scientists have isolated physalolactone C from the leaves, a minor steroidal constituent. A globose, small, 2-locular, light yellowish to orange, aromatic, succulent, edible berry, 1.25 cm to 2.5 cm (1/2 –to 1 inch) in diameter, the size of a cherry tomato, with smooth, glossy, orange-yellow skin and juicy pulp, and like tomatoes, containing many tiny yellowish seeds inside. The berry is encased in the inflated long, papery, tan husk (calyx) which is 3 cm to 3.5 cm in length.

Seeds numerous, pale brown, discoid, 1.75 mm to 2 mm long, minutely shallowly reticulate, embryo curved, endosperm present (Wagner *et al.,* 1999).

The fruits are ripe when they turn yellow-gold. Unripe fruits are green. When they are ripe, the fruit is sweet but with a pleasing grape-like tang. The husk is bitter and inedible. The fruits are covered or encased in a loose, papery husk shaded with purple, which is the persistent calyx and protects them from external injury. They resemble yellow cherry tomatoes. The fruit and husk will naturally dehisce (drop) with a good shake when they are fully ripe or nearly so. They are 5,000 to 8,000 seeds to the ounce (28 g). The fruit of *Physalis minima* is a berry, enclosed within the enlarged, 10-ribbed, reticulately veined calyx, which are 4.1 cm long and 2.5 cm broad. The berries stalked (stalk, 2.2 cm long), almost round having a pinhead-sized depression at the end. They are 1.4 cm to 1.6 cm in diameter, weight 2.15 g and occupy a volume of 1.32 mL. Fruits are light yellow at full maturity. The seeds are globose and yellow. They weigh 1.13 mg each occupy a volume of 1.97 microlitres. The fruits of *Physalis pruinosa* are a lot like those of *Physalis peruviana* in flavor and size, but the husks are tight fitting and they curl back to expose the ripe fruits. The fruits of this plant also drop when they are ripe. The fruits of *Physalis ixocarpa* are shiny, flattened and greenish-yellow purple or reddish when ripe. They range in size from 13 cm to 50 cm (1/2 to 2 inches) in diameter. They are encased in papery husks. The husks split, but do not fall off, as the fruits mature and take on a purplish hue. The flavor of *Physalis ixocarpa* increases with maturity. They are harvested after the husks split. Fruits that feel too hard should be set aside to finish ripening.

Chapter 5

Composition and Uses

5.1 Nutrients

Cape gooseberry fruits are rich in vitamins A, C and B-complex namely (thiamine, niacin, and vitamin B_{12}). The fruit also contains higher amount of vitamin C than the orange, the riper the fruit, the higher the concentration of beta carotene. Cape gooseberries are a good source of vitamin C and provide dietary fiber (Rehm and Espig., 1991), which makes it deal for diets.

Table 5.1: Nutritional Characteristics of Cape gooseberry Fruit. Ranges are presented.

Parameter	Minimum	Maximum
Energy (kJ)	290	290
Water (per cent)	78.9	85.9
Proteins (per cent)	0.5	2.3
Fat (per cent)	0.4	1.3
Carbohydrates (per cent)	11.0	13.3
Fibre (per cent)	2.9	4.9
Ash (per cent)	0.7	1.0
Calcium (mg 100 g^{-1} FW)	7.0	14.0
Phosphorus (mg 100 g^{-1} FW)	21.0	39.0

Contd...

Table 5.1–*Contd...*

Parameter	Minimum	Maximum
Iron (mg 100 g⁻¹ FW)	1.1	1.7
Citric acid (per cent)	1.63	2.30
Malic acid (per cent)	0.25	0.37
Tartaric acid (per cent)	0.18	0.25
Ascorbic acid (mg 100 g⁻¹ FW)	11.0	43.0
Provitamin A (IU)	648	5,000
Thiamine (mg 100 g⁻¹ FW)	0.01	0.10
Riboflavin (mg 100 g⁻¹ FW)	0.04	0.17
Niacin (mg 100 g⁻¹ FW)	0.80	1.73

Sources: Camacho (2000), Durán (2007), Fischer (1995), Hermann (1994), Lieberei and Reisdorff (2007), Rehm and Espig (1991).

5.2 Uses

In the present time, *Physalis peruviana* is employed in herbal medicine systems in Peru, Brazil and Colombia.In Peruvian herbal medicine, this plant is also called *mullaca* or *bolsa mullaca*. There, it is used for asthma, bacterial infections, diabetes, earaches, hepatitis, infection (postpartum), inflammation, itch, jaundice, liver problems, malaria, rheumatism, skin disease, urinary insufficiency, and worms (Wu *et al.,* 2005). In Brazil, *Physalis peruviana* is traditionally used for asthma, blood cleansing, dermatitis, earaches, fever, gallbladder problems, kidney problems, jaundice, liver disorders, malaria, nausea, rheumatism, skin diseases and urinary insufficiency. In Central America, *Physalis peruviana* is used for fever, gonorrhea, malaria, skin diseases and to prevent miscarriages. In Colombia, *Physalis peruviana* is used for asthma, bacterial infections, inflammation, and skin diseases. In Trinidad, this plant species is used for bacterial infections, fever, indigestion, nephritis, and rectitis. In Suriname, *Physalis peruviana* is used for gonorrhea, jaundice, malaria, nephritis, and urinary insufficiency. In Japan, *Physalis peruviana* is used for colds, fever, strep throat, swelling, and urinary insufficiency. In Taiwan, *Physalis peruviana* is used for cancer, fever, hepatitis, liver disease, tumors and urinary insufficiency. In General, *Physalis peruviana* is used for asthma, bacterial infections, boils, cancer, childbirth, dermatosis, diabetes, diarrhea, diuretic, edema, expectorant, eye infections, fainting, fevers, haemostatic, hemorrhage (postpartum), infertility, inflammation, leukemia, malaria, nausea, pain, tumor (testicle), skin disease, sleeping sickness, stomach problems and as an antiseptic. In addition to *Physalis peruviana's* anticancerous and antileukemic actions, several research groups have confirmed *Physalis peruviana's* antibacterial and antiviral activity.

P. peruviana is widely used in folk medicine for treating diseases such as malaria, asthma, hepatitis, dermatitis, diuretic and rheumatism (Wu *et al.,* 2005). Its extract can also be used for preparing health drinks. Ahmad *et al.* (1999) reported that the plant is diuretic and the juice of its leaves is given for worm and bowel complaints, while heated leaves are applied as a poultice. An extract of the leaves shows antibiotic activity against Staphylococcus (Perry and Metzger, 1980). The extracts of different parts of the plant show anti-hepatotoxic (Arun and Asha, 2007) and anti-proliferative effects on hepatoma cells (Wu *et al.,* 2004). A supercritical carbon dioxide extract of cape gooseberry leaves exhibits enhanced antioxidant and anti-inflammatory activities (Wu *et al.,* 2006). Also extracts of calices confirmed the anti-inflammatory activity and validated its use in folk medicine (Franco *et al.,* 2007). Ingestion of cape gooseberry fruits decreases glycemia (Rodríguez and Rodríguez, 2007). The high β carotene content of the cape gooseberry fruit has a potentially anticarcinogenic effect (Steinmetz and Potter, 1996). Leaf extract of this species exhibited moderate *in vitro* anticancer (breast, renal and melanoma) actviity (Fouche *et al.,* 2008).

Herbal practitioners in both South and North America today rely on *Physalis peruviana* for various bacterial and viral infections as well as a completely therapy for cancer and leukemia. Although not widely available in the United States. *Physalis peruviana* is found as an ingredient in various herbal formulas and in bulk supplies. In the United States, *Physalis peruviana* is hardly seen, except in Pennsylvania (Dutch county and parts of the Midwest). Interestingly enough, much of the clinical research has ignored the local and indigenous uses of the plant; thus, many of its effective uses in herbal medicine remain unexplained. Its tested antibacterial properties could validate its use as an antiseptic and disinfectant for skin diseases and its use to treat gonorrhea. The fruits of *Physalis peruviana* are juicy, mildly astringent and sweet with a pleasant blend of acid. Due to its high content of pectin and acidity, the fruits are also used in making excellent jam, jellies and sweet pickle (Mazumdar, 1979). Cooking the juice with sugar makes thick cape gooseberry-flavored syrup. Also cape gooseberries serve in the preparation of sauces, which are then used in deserts as a flavoring for cakes. But also, ripe berries are used in enriched preparations of meat and curries (Mazumdar, 1979). The overall quality of the fruit is good can be used in order to prepare desserts (pies, cakes, jellies, compotes, jams etc.). Its flavor has been defined as a pleasant, unique tomato/pineapple like blend. The dried fruit can be used as a raisin substitute, though it is not so sweet. The dried fruit is also said to be a substitute for yeast. The citrus fruits, such as oranges, lemons, grapefruits, tangerines and limes, contain good quantities of vitamin C but little carotene. In contrast, papayas, mangoes and cape gooseberries (*Physalis peruviana*) contain both carotene and vitamin C. They are eaten and liked by all. They are juicy and, as it evident from their chemical composition, they are a good source of vitamin C. The raw fruit can also be used as vegetable. Their uses are similar to common tomato. They can be eaten raw, used in salads, desserts, as a flavoring, and in jams and jellies. These fruits are excellent when dipped in chocolate, and can be dried and eaten. Although the plant conveniently wraps up each fruit in its own 'paper bag' (botanically, the calyx) to protect it from pests and the elements, this calyx is toxic and should not be eaten. *Physalis peruviana* keeps well and makes excellent preserves. The canned fruits have been exported from South Africa and the jam from England. In Peru and Chile, these fruits are casually eaten and occasionally sold in markets but the plant is still not an important crop. In the 18th century, the fruits were perfumed and worn for adornment by native women in Peru.

5.3 Medicinal Properties

Physalis peruviana L. is Used As

- ☆ Analgesic (pain-reliever)
- ☆ Anti-asthmatic
- ☆ Antibacterial
- ☆ Anticancerous
- ☆ Anticoagulant
- ☆ Anti-diabetic

☆ Anti-emetic

☆ Antihemorrhagic (reduces bleeding)

☆ Anti-inflammatory

☆ Antileukemic

☆ Antimalarial

☆ Antimicrobial

☆ Antimycoplasmal

☆ Antipyretic

☆ Anti-rheumatic

☆ Antiseptic

☆ Antispasmodic

☆ Antitumorous

☆ Antiviral

☆ Appetizing

☆ Astringent

☆ Blood cleanser

☆ Blood thinner

☆ Disinfectant (postpartum infections)

☆ Disinfectant (skin diseases)

☆ Diuretic

☆ Expectorant

☆ Febrifuge (reduces fever)

☆ Food

☆ Hepatic

☆ Hepatotonic (tones, balances, strengthens the liver)

☆ Hypoglycemic

☆ Hypotensive (lowers blood pressure)

☆ Immunomodulator (modulates some overactive immune cells)

☆ Immunostimulant

☆ Laxative

☆ Narcotic

☆ Nephritic

☆ Prnamental

☆ Purgative renal

☆ Sedative

☆ Tonic

☆ Vermifuge (expels worms)

☆ Enema

☆ Source of vitamin C (raw fruit)

Physalis peruviana L. is Used for/Against

☆ Abdominal pains

☆ Asthma

☆ Bacteria

☆ Bacterial infections (all kinds)

☆ Blood pressure

☆ Bowel complaints

☆ Cancer coagulation

☆ Dermatitis

☆ Diabetes

☆ Earache

☆ Female disorders

☆ Fever

☆ Gallbladder problems

☆ Hepatitis

☆ Hypertension (high blood pressure)

☆ Infertility

☆ Infections

☆ Inflammations (skin)

☆ Jaundice

☆ Kidney problems

☆ Leukemia

☆ Liver problems

☆ Malaria

☆ Microbes

☆ Miscarriage (preventive)

☆ Mycobacteria infections

☆ Mycoplasma infections

☆ Mycopartum infections

☆ Psoriases

☆ Rheumatism

☆ Rosaceous

☆ Scleroderma

☆ Skin diseases

☆ Skin infections

☆ Spasms

☆ Spleen

☆ Viral infections (all kind)

☆ Virus

☆ Vomiting

☆ Worms

Cancer - Anticancerous

Physalis peruviana has been the subject of recent clinical research (which is still ongoing), based on the preliminary studies showing that it is toxic to numerous types of cancer and leukemia cells. The new steroids found in *Physalis peruviana* have received the most attention, since many of the documented anti-cancerous, anti-tumorous and anti-leukemic actions are attributed to these steroids. Various extracts of *Physalis peruviana,* as well as some novel plant steroids called physalins, have shown strong *in vitro* and *in vivo* (mice) activity against numerous types of human and animal cancer cells including lung, colon, nasopharynx, liver, cervix, melanoma and giloma and glioma (brain) cancer cells. This cancer research began in the early 1980s with researchers in Thailand and the United States and was verified with research performed at the University of Taiwan in 1992 (where they demonstrated a significant effect against five human cancer cell lines and three animal cancer cell lines). Then in 2001, researchers at the University of Houston isolated yet another new chemical in *Physalis peruviana* which demonstrated remarkable toxicity against nasopharynx cancer cells, lung (adenocarcinoma) cancer cells as well as leukemia in mice. The same Taiwanese researchers had already published a separate study on *Physalis peruviana's* other anti-leukemic phytochemicals in 1992, reporting that two physalin chemicals inhibited the growth of five types of acute leukemia, including lymphoid (T and B), promyelocytic, myeloid and monocytic. With tumor cells, research suggests that several of the steroidal chemicals in *Physalis peruviana* act on an enzyme level to arrest the normal cell cycle in cancer cell as well as cause DNA damage inside of cancer cells (making them unable to replicate).

Diabetes – Antidiabetic

To treat diabetes, the roots of three *Physalis peruviana* plants are sliced and macerated in ¼ liter of rum for seven days. Honey is added, and ½ glass of this medicine is taken twice daily for 60 days. Western scientists did somewhat validate the indigenous use for diabetes when they reported a mild hypoglycemic effect in mice fed a water extract of the root. One must wonder the result would have been if they has followed native customs and employed an alcohol extract instead. It is said in Mexico that a decoction of the calyces of *Physalis ixocarpa* Brot. ex. Hornem. Will cure diabetes.

Diuretic

Indigenous tribes in the Amazon use an infusion of the leaves of *Physalis peruviana* as a diuretic. In Brazilian herbal medicine the plant is employed as a diuretic. It is said to be a good diuretic. In Colombia, the leaf decoction is also taken as a diuretic. Kirtikar and Basu (1935) have reported that the plants *Physalis minima* are diuretic. The fruits of this closely related plant species are also considered to be diuretic in the Punjab a region in India.

Earache

The leaves and/or roots of *Physalis peruviana* are used in Peru for earache. In the Brazilian Amazon, indigenous tribes use the sap of the plant for earaches.

The mundas, a tribe of Chhota Nagpur, India, mix the juice of the leaves of *Physalis minima* with water and mustard oil and use it as a remedy against earache.

Female Disorders

Physalis peruviana is used by indigenous peoples for female disorders.

Infertility

In the Solomon Islands, the fruit of *Physalis peruviana* is decocted and taken internally to promote fertility.

Miscarriage

A tea is made of the entirely plant and/or the leaves in the West Indies and Jamaica to prevent miscarriage.

Postpartum Infections

In Peru the leaf is infused and used to treat postpartum infections. Possibly, the antispasmodic and muscle contractive properties documented for *Physalis peruviana* might explain its widespread use for female disorders.

Fever – Antipyretic

In Brazilian herbal medicine the plant is employed for fever. For the same purpose, *Physalis peruviana* is used in other parts throughout the rainforests, although its widespread use in the rainforests for fevers remains unexplained by science. The ripe fruits are considered a good source of Vitamin P and are rich in pectin. The fruit is also rich in vitamin A, vitamin C and some of the B complex (thiamine, niacin and B12). The protein and phosphorus levels are exceptionally high for a fruit. Other analysis determined 3000 I.U. of carotene per 100 g (vitamin A). Be careful. Fruits seen on the market vary in taste and size. There is great genetic variability.

Gallbladder Problems

In Brazilian herbal medicine the plant is employed for many types of gallbladder problems.

Hepatitis

The leaves and/or roots are used in Peru for hepatitis. The roots and/or leaves are prepared as an infusion. Its antiviral properties could well explain its long history of use for hepatitis, although scientists have not tested it especially against hepatitis.

Immune Stimulant

Physalis peruviana has been the subject of recent clinical research (which is still ongoing), based on the preliminary studies showing that it is an effective immune stimulant. The new steroids foods in *Physalis peruviana* have received the most attention, since many of the documented immune stimulant actions are attributed to these steroids. Researchers in China and Russia independently demonstrated significant immunomodulatory effects against blastogenesis (a process triggered in leukemia) while boosting other immune functions which might account for the anti-leukemic effects in mice seen by other researchers.

Jaundice

Indigenous tribes in the Brazilian Amazon use of the plant for jaundice.

Kidney Problems

In Brazilian herbal medicine the plant is employed for many types of kidney problems.

Laxative

Kirtikar and Basu (1935) have reported that the plants of *Physalis minima* are laxative.

Liver Problems

In Brazilian herbal medicine the plant is employed for many types of liver problems. The leaves and/or roots are used in Peru for liver problems.

Malaria – Antimalarial

The leaves and/or roots are used in Peru for malaria. The dosage is 1 cup of tea made from the aerial parts of the plant. *Physalis peruviana* is also used for this purpose in many other parts of the rainforest, although its widespread use throughout the rainforests for malaria remains unexplained by science. In Ghana, an herbal medication against malaria s being studied. This medication consists of a decoction prepared with *Jatropha curcas, Gossypium hirsutum, Physalis angulata* and *Delonix regia.*

Microbes – Antimicrobial

Physalis peruviana has been the subject of recent clinical research (which is still ongoing), based on the preliminary studies showing that it has antimicrobial properties.

Virus – Antiviral

Several Research Groups have confirmed *Physalis peruviana*'s antiviral activity. Research groups in Japan, for example, have been focusing on *Physalis peruviana*'s antiviral actions and preliminary studies show that it is active *in vitro* against Polio virus I, Herpes simplex virus I, the measles virus, and HIV-I, demonstrating reverse transcriptase inhibitory effects. Its antiviral properties could well explain its long history of use for hepatitis, although scientists have not tested it specifically against hepatitis.

Bacteria –Antibacterial

Recently, in 2000 and 2002, *Physalis peruviana* was show to be active *in vitro* against several strains of mycobacteriums and mycoplasmas (both very stubborn types of bacteria which are not widely susceptible to standard antibiotics).

In addition to these actions, *Physalis peruviana* has demonstrated effective antibacterial properties *in vitro* against numerous types of gram positive and gram negative bacteria, including *Pseudomonas, Staphylococcus* and *Streptococcus.*

Purgative

In the Punjab, the fruit of *Physalis minima* is considered to be a purgative.

Rheumatism

The leaves and/or roots are used in Peru for rheumatism. In Brazilian herbal medicine the plant is employed for chronic rheumatism.

Sedative

In Brazilian herbal medicine the plant is employed for skin diseases and dermatitis. Folk medicine also recommends it for kin diseases such as dermatitis, psoriasis, skin infections, rosaceae, scleroderma, etc.

Spasms –Antispasmodic

Physalis peruviana has also been reported to reduce spasms in guinea pigs and to contract isotonic muscles in toads. Possibly, the antispasmodic and muscle contractive properties documented for *Physalis peruviana* might explain its widespread use for asthma and female disorders as well.

Spleen

Kirtikar and Basu (1935) have reported that the plants of *Physalis minima* are useful in enlargement of the spleen.

Tonic

Kirtikar and Basu (1935) have reported that the plants of *Physalis minima* are used as a tonic. In the Punjab, India, the fruit is considered to be a tonic, as well.

Vomit

In Brazilian herbal medicine the plant is employed for vomiting.

Worms – Vermifuge

Indigenous peoples in the Peruvian Amazon use the leaf juice of *Physalis peruviana* internally and externally for worms.

Dessert quality

The fruits are juicy, mildly astringent and sweet with a pleasant blend of acid. The overall quality of the fruit is good.

Precautions

- ☆ This plant species is not a substitute for standard medical treatments.
- ☆ Any plant substance can cause an allergic reaction in some people.
- ☆ Do not try to self diagnose or attempt treatment for any serious or long term problem or while undergoing a prescribed course of treatment without consulting a medical professional.
- ☆ All parts of the plant, except the fruit, are said to be poisonous (Stary, 1983; Frohne and Pfander, 1984).
- ☆ Excessive ingestion of fruit of *Physalis peruviana* may thin the blood and lower blood pressure.

Ornamental

The most popular *Physalis* plant species grown as ornamentals are *Physalis alkekengii, Physalis franchettii, Physalis bunyardii, Physalis monstrosa* and *Physalis nana*. *Physalis alkekengii,* a closely related species, and its varieties are grown for the decorative value of their brilliantly colored, swollen calyces. These resembles miniature "Chinese Lanterns", thus one of the common names, Chinese Lantern Plant. Other popular named include Alkekengi and Winter Cherry. In gardens, they are sometimes grown as annuals. The Chinese Lantern Plant and its varieties are very pretty in the garden in late summer. The stems bearing the large orange "fruits" are valued for cutting and bringing indoors during the fall and winter months. They last a long time and look quite handsome in a vase. "Fruits" refers to the swollen calyces of the white flowers, which open early in the summer. If they are to be brought in to decorate the house, the stems bearing the pretty calyces should be cut as soon as they have developed their full coloring.

Chapter 6

Cultural Practices

6.1 Soil

Physalis species thrive in most soil types and will do very well in poor soils and in pots. They are easily grown in any well-drained soil in a position full of sun but do best on sandy to gravelly loam. They will not flourish in heavy, poorly drained ground. Light soil is the best for their development. Heavy ground can be enhanced, however, by digging deeply and adding sand and compost. On highly fertile alluvial soil, there is much vegetative growth and the fruits fail to color properly. Very good crops are obtained on rather poor sandy ground. Where drainage is a problem, the plantings should be on gently slopes or the rows should be mounded. The plants become dormant in drought. *Physalis peruviana*, should be grown in loamy soil. If the ground is full of clay, they should be planted in raised beds. Set them 30 cm apart. Use black plastic mulch. Do not dig manure or mulch into the soil fertilize heavily; that will cause very lush growth at the expense of the fruits.

The ability of *Physalis peruviana* to germinate and to emerge under low saline stress conditions could indicate that the species posses the certain salt tolerance genetic potential, atleast during its early development stage. This does not necessarily indicate that plantlets indicated in saline stress conditions to grow and complete their adult plant life cycle in these circumstances (Miranda *et al.,* 2010). However studies carried out by Ulloa *et al.* (2006) on adult cape gooseberry plants exposed to increasing NaCl concentrations have demonstrated that this species is moderately tolerant to salinity as it manages to perform well even at conductivity value up to 6 dsm$^{-1.}$

6.2 pH

The plant prefers acid, neutral and basic (alkaline) soils and can grow in very acid and very alkaline soils. Plants tolerate a pH in the range 4.5 to 8.2.

6.3 Climate

Physalis peruviana enjoys fairly warm (but not hot) temperatures, and protection from frost. *Physalis peruviana* is an annual in temperate regions and a perennial in the tropics. In areas where frost or freezes occur, plants are easily grown as annuals. Provide lots of water throughout the growing year, except towards fruit ripening time. In England, the plans have been undamaged by 3 degrees of frost. In South Africa, plants have been killed to the ground and failed to recover after a temperature drop to 30.5°F (−0.75°C). *Physalis ixocarpa*, a closely related species native to Mexico, prefers warm, dry climates. They do not last long in humid, rainy climates. *Physalis ixocarpa* also requires a longer growing season and is renderer than the other.

6.4 Cultivars/Genotypes, Genetic Variability and Improvement

Most of the research on Cape gooseberry deals with the development and improvement of growing techniques (Klinac, 1986, Wolf 1991 and Chattopadhyay, 1996).

There is a lack of breeding efforts; only some research related to selection among the different accessions or to the development of *in vitro* culture protocols to obtain soma clonal variation or as a first step for genetic transformation has been (Torres *et al.,* 1995) conducted. Genetic difference in yield and fruit quality characters found among accessions from different regions (Prohens and Nuez, 1994) can be exploited for cape gooseberry breeding. A lot of variability in terms of vegetative growth, fruit size, quality, composition and colour among different genotypes/

cultivars were reported (Singh *et al.,* 2011: Singh *et al.,* 2012, Marquez *et al.,* 2009:. Evaluation and screening programme of CITH Srinagar J and K (Singh *et al.,* 2011: Anonymous 2011-12, Anonymous 2012-13) production, fruit quality, colour characters sketch four genotypes (CITH CGB S20- CITH CGB S3- CITH CGB S12- CITH CGB S1- CITH CGB S6) of cape gooseberry based on production and quality attributes suitable for temperate regions (Tables 6.1 and 6.2).

Table 6.1: Physical Characters of different Varietal Genotypes/Selections of Cape gooseberry Grown Under Temperate Climate of Kashmir, India

Genotypes	Characters						
	Plant Height (cm)	No. of Fruits/ Plant	Yield/ Plant (g)	Fruit Weight (g)	Fruit Length (mm)	Fruit D (mm)	Firmness (RI)
CITH CGB S1	62.69	40.00	606.26	15.15	27.17	31.22	38.55
CITH CGB S2	100.32	57.66	825.58	14.20	23.63	33.35	49.57
CITH CGB S3	46.24	36.33	686.83	19.10	29.82	34.19	39.59
CITH CGB S4	76.37	32.00	199.67	6.51	20.22	24.49	63.30
CITH CGB S5	98.56	57.33	548.57	9.27	20.42	26.63	53.65
CITH CGB S6	46.32	61.00	496.53	8.14	22.10	24.35	39.28
CITH CGB S7	73.71	38.66	386.44	9.86	22.13	26.36	59.70
CITH CGB S8	62.67	45.66	464.33	10.17	23.07	27.02	63.34
CITH CGB S9	50.35	67.66	924.81	13.52	24.16	28.98	32.13
CITH CGB S10	63.54	64.00	467.06	7.40	22.22	23.56	61.70
CITH CGB S11	62.75	60.33	484.02	7.90	20.79	24.22	56.54
CITH CGB S12	45.51	42.66	715.67	16.44	28.84	34.22	41.69
CITH CGB S13	114.67	26.00	347.86	13.47	25.42	27.63	64.33
CITH CGB S14	49.08	60.66	428.71	7.03	22.54	23.42	56.34
CITH CGB S15	107.73	36.00	337.99	9.37	22.28	25.36	57.29
CITH CGB S16	63.53	73.00	552.95	7.63	22.25	23.06	53.51
CITH CGB S17	43.47	62.00	684.45	11.29	21.71	28.47	61.60
CITH CGB S18	73.42	58.00	511.58	8.72	22.36	24.72	60.18
CITH CGB S19	63.40	63.00	721.08	11.51	23.42	27.69	62.18
CITH CGB S20	50.12	72.66	1145.03	15.33	28.45	31.51	62.66
CD	0.953*	3.818*	3.83*	1.35*	1.79*	1.08*	0.59*

Singh *et al.,* 2011: Anonymous 2011-12.

Table 6.2: Quality Attributes of Different Genotypes/Selections of Cape gooseberry Grown Under Temperate Climate of Kashmir, India

Genotypes	Quality Attributes				
	Juice (per cent)	TSS (°B)	Total Titrable Acidity (per cent)	Ascorbic Acid (mg/100g)	Carotene (mg/100g)
CITH CGB S1	60.97	9.71	0.53	19.57	1.57
CITH CGB S2	59.41	8.16	0.28	22.72	1.63
CITH CGB S3	62.17	8.76	0.47	21.24	1.56
CITH CGB S4	61.19	8.62	0.52	21.31	1.66
CITH CGB S5	59.35	9.14	0.47	19.54	1.62
CITH CGB S6	61.33	9.25	0.56	24.31	1.67
CITH CGB S7	60.51	9.06	0.61	22.68	1.64
CITHC GB S8	58.2	9.3	0.56	22.71	1.59
CITH CGB S9	61.1	8.77	0.51	22.36	1.46
CITH CGB S10	61.67	6.92	0.44	19.61	1.72
CITH CGB S11	60.13	8.75	0.53	21.21	1.58
CITH CGB S12	62.77	8.43	0.58	22.68	1.56
CITH CGB S13	61.32	8.3	0.58	19.63	1.48
CITH CGB S14	63.17	7.29	0.47	21.2	1.61
CITH CGB S15	58.14	8.15	0.58	22.75	1.52
CITH CGB S16	58.23	9.28	0.53	21.23	1.47
CITH CGB S17	56.2	9.13	0.47	22.66	1.58
CITH CGB S18	58.43	9.22	0.44	19.55	1.55
CITH CGB S19	57.01	8.11	0.67	22.79	1.59
CITH CGB S20	61.07	7.31	0.42	19.55	1.63
CD	1.90*	0.29*	0.011*	0.30*	N.S.

Singh *et al.*, 2011: Anonymous 2011-12.

A simple breeding strategy in the development of the hybrids to exploit heterosis for yield characters (Basra, 1999). The development of hybrids can be a good strategy for breeding of this crop, especially when looking for good adaptation to glass house cultivation. Moreover cape gooseberry hybrid seed is easy to produce because the flowers has a low sensitivity to manipulation and considerable number of seeds (>300) per fruit are usually obtained from each crossing. The use of hybrids can improve the yield performance of cape gooseberry without effecting fruit quality (Leiva Brondo *et al.,* 2001).

Ligarreto *et al.* (2005) named various countries in Latin America and the Caribbean (Brazil, Chile, Colombia, Costa Rica, Ecuador, *Guadalupe*, Guatemala, Mexico and Peru), where collections of *Physalis* germplasm exist, mostly in universities and/or research stations. For example, in Colombia, 222 holdings of Physalis are located in the National University (Palmira) and 98 accessions in the Corpoica research stations: Rionegro and Tibaitata. The genetic variability of this genus in Latin America and the Caribbean is represented by traditional varieties, mostly wild ones, except in Brazil and Mexico, where the improved varieties of *P. peruviana* and *P. philadelhica* exist (Ligarreto *et al.,* 2005). The National Research Council (1989) suggests selecting cape gooseberry plants with superior and sweet fruit types and with a uniform growth habit along the Andean cordillera where the greatest variation is likely to be found. The Council also recommends selecting plants whose shapes are amenable to mechanical harvesting which would also be a major advantage, provided that the fruit mature uniformly, all at once. Verhoeven (1991) reported that in Australia the cape gooseberry is commercialized under the names of the cultivars such as 'Golden Nugget' or 'New Sugar Giant', that develop large fruits but with an incipient flavor. But Australian fruits with a smaller size present a better flavor and are preferred for elaboration of marmalades and other processed foods. Wolff (1991) reported three commercially important varieties from Baton Rouge (USA), 'Peace', 'Giant Groundcherry' and 'Goldenberry'. Five varieties were described by CRFG (1997), 'Giallo Grosso', 'Giant', 'Giant Poha Berry', 'Golden Berry' and 'Long Ashton Golden Berry', all with fruits approximately 1 inch in diameter. At Calcutta University in India, cv. Rashbori is used in testing programs (Majumdar and Mazumdar, 2002). From Germany, G. Fischer introduced two ecotypes 'Kenya' and 'South Africa' to Colombia. These African cape gooseberries bear larger fruits. Fischer (1995) found Kenyan and South African cultivars with an average fresh weight of 6.2 g and 6.7 g, respectively, compared to the Colombian provenience with 4.2 g. The African ecotypes possessed a higher provitamin A content (Fischer *et al.,* 2000), but a more intense yellowish-orange color and higher total soluble solids (TSS) content favored the Colombian fruits (Fischer, 1995). Different cape gooseberry ecotypes can present wide variations in chromosome number. Rodriguez (2004) found 2 n = 24 in wild ecotypes, but 2 n = 32 in 'Colombia' and 2 n = 48 in 'Kenya' provenience. The National Research Council (1989) indicated that this interesting botanical relative of tomatoes and potatoes has commercial promise for many regions, and recommended this species to tomato breeders seeking a challenge.

Common cultivars grown in the world are:

Giallo Grosso

The large golden fruit is eaten raw or preserved after ripening. In areas with mild winters the plant will last for several years.

Giant

Large, golden-orange fruit, approximately 1 inch in diameter with a delicious flavor. Vigorous, spreading plants grow 3 to 5 feet tall. Requires a long growing season.

Giant Poha Berry

Fruit is approximately 1 inch. The leaves are fuzzy, green-grey and different from other Physalis. Plant grows from 1 to 2-1/2 feet tall.

Golden Berry

Fruits average 1 inch in diameter, with some reaching 2 inches. Pulp is very flavorful and sweet. Deseeded fruit juice similar in color and intensity of taste to orange juice. Dried fruits are used in fruit cakes in place of raisins. Said to be resistant to light frosts which have caused tomatoes and other *Physalis* species and cultivars to die.

Golden Berry, Long Aston

Original Long Ashton selection of Golden Berry. Rich golden fruit, said to be superior to other types.

Other cultivars mentioned in various sources include Dixon, Garrison's Pineapple Flavor, New Zealand, Peace and Yellow Improved.

Cape gooseberry is usually grown out doors so there has ben no selection under glass house conditions. Therefore this crop probably presents little adaptation to these conditions. It has been suggested that heterozygotes have an adaptative adventage when they are moved to

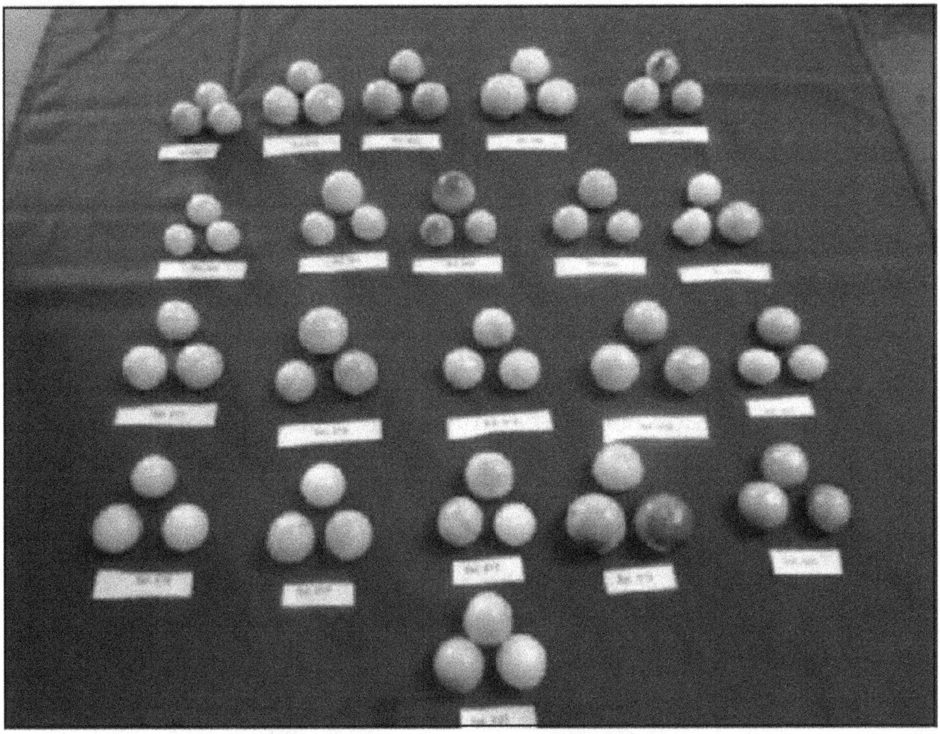

Figure 6.1: Variability of Cape gooseberry Fruits.

new conditions or are placed under stress, because their interaloculus variation can result in a higher developmental homeostasis than that of homozygotes (Lerner, 1954, Blum, 1988, Nuez *et al.,* 1997: Kang, 1998) which could explain the better behavior of hybrids grown in the glass house. Screening of genotype/germ-plasm adapted to green house conditions can lead to the diversity of interesting materials for protects cultivation if accessions coming from cloudy environment are tried. Regarding fruit quality screening of collections for fruit weight and SSC can be exploited for identification of source for variation. In addition development of the hybrid heterosis to yield fruits can not lead only to higher yield but also big fruits. The development of hybrids can be a good strategy of breeding for the crop, especially when looking for good adaptation to glass hybrids, seed is easy to produce, because the flower has a low sensitivity to manipulation and considerable numbers of rows (> 300) for fruit are usually obtained from each colny. These hybrids can improve the yield progress of Cape gooseberry without affecting fruit quality (Leiva Brondo, 2001).

6.5 Season

The tropical season and subtropical parts of India the fruit ratio in Feb. to March and in temperate regions seeds are sown in spring (April month) and begin to fruit in August and continue until there is a stormy frost.

6.6 Propagation

The easiest way to increase these plants is to lift and separate them into rooted pieces in the spring and replant them. They may also be started from seeds, which are sown in flats of sandy soil or directly outside. Those sown directly outside should be thinned to 6 inches apart before they become too crowded. *Physalis peruviana* propagates easily from the many seeds the fruit contains. There are 5,000 to 8,000 seeds to the 28-30 g of fruit and, since germination rate is low, this amount is needed to raise enough plants for an acre and 70 g for a hectare. In India, the seeds are mixed with wood ash or pulverized soil for uniform sowing. Sometimes propagation is done by means of 1-year-old stem cutting treated with hormones to promote rooting, and 37.7 per cent success has been achieved. The plants thus grown flower early and yield well but are less vigorous than seedlings. Air-layering is also successful but not often practiced.

In India, seeds are broadcast from March through May. In Hong Kong, planting in seedbeds is done in September/October and again in March/April. In the Bahamas the first seeds planted in late summer of 1992 produced healthy plants and a continuous crop of fruits for 3 months during the following winter.

In Jamaica, the initial planting of *Physalis peruviana* in late January made slow growth until June when development accelerated. By mid-August the plants reached 15 inchs (37.5 cm) in height with much lateral growth, and were flowering and setting fruit. It would appear that the heat of summer is unfavorable for fruit development and therefore, the best time to plant *Physalis peruviana* is in the fall so that fruit can be set during the cooler weather and harvested in late spring or early summer. In California, the plants do not fruit heavily until the second year unless started early in greenhouses.

Some growers have kept plants in production for as long as 4 years by cutting back after each harvest, but these plants have been found more susceptible to pests and diseases. *Physalis alkekengii*, a closely related species, is a perennial spread by means of underground stems. In gardens, they are sometimes grown as annuals.

6.7 Seed Growing

For seed growing fruits should be gathered in late summer once it turns golden yellow and papery husk split open. Fruits are wrapped in sheet cloths, crushed and water is run over the pulverized fruits to since away the pulp. Tiny flat seeds are pickled, stored in an air tight container away from direct light until spring. Seeds are sown in late spring (end April to May) under temperature conditions of Kashmir, India. Seeds are sown to the depth of ¼-inches in 6-inches plastic pots filled with 3 parts potting soil and 1 part perlite or leaf mould and in open rows under protected containers/polyhouses. Water each bed/pot using spray bottle or garden hose with a mist nozzle attachment. Soil is kept mist to a depth of 1-inch at all time. For sprouting germination of seeds 8-10 days are generally required. The plants are kept in pots/soil holes until soil temperature reach 8-20°C. The seedlings are field planted when they are 15 cm to 20 cm tall with at least 1.0 m between each plant.

6.8 Cutting Growing

4-6-inches long cuttings from a healthy mature cape fruit plant in spring before plant forms flower bud. Severe the cutting just below a set of leaves using a pain of sterilizer floral strips. Strip off the bottom 2 leaves to expure the intersoils. Dip the defoliated end of Cape gooseberry cutting into 0.2 percent IBA rooting hormones. Stick the Cape gooseberry cuttings up to its lowest set of leaves into 6" pot fitter with perlite or soil; leaf mixture (1:1). Spritz the perlite with a spray bottle until water begins to trickle from the drain holes at the bottom of the pot. Place the potted cape gooseberry cutting near a source of strong yet diffuse light. Keep temperatures surrounding the pot around 20-22 °C. Mist the foliage once or twice a day to keep the leaves from drying out on the edges. Check for roots in 21 to 30 days by gently pushing aside some of the perlite to reveal the internodes. Look for spindly white roots emerging from the internodes. Transplant the rooted cape gooseberry cutting into a 6-inch plastic pot filled with garden soil. Place it outdoors under partial shade for two weeks to help the plant acclimate to outdoor conditions. Plant the cape gooseberry plant in a permanent bed with slightly sandy soil once it begins to put on significant height and foliage growth.

6.9 Spacing

In India, plants of *Physalis peruviana* 15 cm to 20 cm high are set out 45 cm apart in rows 0.9 m apart. Farmers in South Africa space the plants 0.6 m to 0.9 m apart in rows 1.2 m to 1.8 m or even 2.4 m apart in very rich soil under plastic green house condition spacing of these rows at 1.2m, 0.8m wherein single lines and 0.8m between plants with 12500 plants/hectare with a yield of 22.4t/ha (Ayala, 1990).

6.10 Pruning

Very little pruning is needed unless the plant is being trained to a trellis. Pinching back of the growing shoots will induce more compact and shorter plants.

6.11 Frost Protection

In areas where frost may be a problem, providing the plant with some overhead protection or planting them next to a wall or a building may be sufficient protection. Individual plants are small enough to be fairly easily covered during cold snaps by placing plastic sheeting, etc. over a frame around them. Plastic row covers will also provide some frost protection for larger plantings. Potted specimens can be moved to a frost-secure area. Young growth at the ends of the branches is particularly susceptible to frost damage.

Plant in a location that enjoys **full sun** and remember to **apply water fairly sparingly**. Cape gooseberry is generally regarded as a **tender** plant, so remember to ensure that temperatures are mild before moving outdoors.

6.12 Light

Physalis species are easily grown in a position full of sun. They will not flourish in a shady place. This plant species needs full sun but protection from strong winds.

6.13 Fertilizing

The plants appear to need little or no fertilizer. Fruit production decreased significantly when fertilizer was applied. Under heavy fertilizing, the plants exhibited a great deal of vegetative growth but produced few flowers or fruit. Plants planted in sandy soil without any amendment or fertilizer produced 150 to 300 flowers per plant with a corresponding number of fruit.

6.14 Manures and fertilizers

F.Y.M.	30-40 t/ha	Entire FYM, P and K
Nitrogen (N) kg/ha	120-150 kg/ha	along with ½ N as
Phosphorus (P)	90-120 kg/ha	basal dose and ½ N
Potash (K)	60-90 kg/ha	as top dressing 30-40days after transplanting

Chapter 7

Growth and Development

7.1. Flowering and Pollination

In the Cundinamarca state of Colombia, cape gooseberry flower bud development lasted between 18 and 21 days (Mazorra *et al.,* 2006). In India, flowering initiated 70 to 80 days after transplanting and 19 to 23 days passed between the flower buds' initiation and anthesis (Gupta and Roy, 1981). These authors observed that during flowering (3 to 4 days) the corolla opened in the morning and closed in the evening. Lagos *et al.* (2008) found in the Nariño state of Colombia that the corolla opened between 7:00 and 10:00 h and closed between 16:00 and 18:00 h with petal fall 5 to 6 days after the first floral opening.

The yellow, bell shaped flowers are easily pollinated by insects and wind (National Research Council, 1989), self-pollination is common (Gupta and Roy, 1981). The majority of fruits (85 percent of set fruits) developed under open pollination (non-artificial) (Gupta and Roy, 1981). Lagos *et al.* (2008) observed that 2 days before flower opening, pollen matured and stigma was receptive, a phenomenon that restricts auto-pollination. In addition, they observed mixed pollination with 54 percent of cross-pollination in *P. peruviana.*

The Cape gooseberry flowers year-round in frost-free areas (National Research Council, 1989). The first flowers are often sacrificed to ensure the establishment of strong, healthy plants. Also, removing the first flowers avoids the fruit cracking when few but big fruits are formed in the lower part of the plant (Fischer, 2005).

7.2. Fruit Growth, Developmnet and Maturation

When fecundation occurs, the ovary has a length of 2.0 to 2.5 mm. At this moment the corolla abscises and fruit begins to grow, simultaneously with the calyx (Valencia, 1985; Mazorra *et al.,* 2006). The calyx grows at a faster rate than the ovary (Yamaguchi, 1983) and at the ripening stage the husk of the Colombian ecotype is about 2.5 times the length of the fruit (Valencia, 1985).

Fruit development in field trials in India lasted 50 days (Gupta and Roy, 1981) and 60 days for the Rashbhori cultivar (Majumdar and Mazumdar, 2002), in Germany 56 to 63 days (Wonneberger, 1985), in France 70 days (Peron *et al.,* 1989) and in Colombia between 60 and 80 days, depending on the agroecological site conditions (Galvis *et al.,* 2005). Thus, tropical altitude influences duration of fruit development. In the Boyacá state of Colombia (4°N) at 2,690 m a.s.l. (12.5°C mean temperature), cape gooseberry fruits required 75 days to harvest, whereas at a lower altitude, 2,300 m (17.0°C), development was faster, only 66 days (Fischer *et al.,* 2007). From New Zealand, Klinac (1986) reported a faster fruit development when rooted cuttings were used, rather than plants originated from seeds.

The size and weight growth of the fruit show a typical sigmoid curve. Whereas the fruit increases its size constantly during the 60 days of its development, the calyx stops its expansion 20 to 25 days after fruit set (Fischer, 2000). The fruit tends to grow more in longitude between days 10 and 25, contrary to the following days during its maturation when it grows more in the transversal diameter (Fischer *et al.,* 1997a; Fischer, 1995). Also, fruits developed on the leaf axils of the main stems are slightly heavier than those from the lateral shoots (Mazorra *et al.,* 2003).

The Cape gooseberry fruits possess the capacity to accumulate high amounts of water and sucrose until the consumer maturity stage (orange skin color), assuming a water supply up to the last moment before harvest, in detriment of postharvest quality and longevity (Fischer *et al.,* 1997b; Fischer and Martínez, 1999). During fruit development, carbohydrate patterns are similar between fruit and calyx, which confirms the close relation in carbohydrate metabolism between these two organs (Fischer and Lüdders, 1997), but anatomically the husk is more leaf-like (Fischer *et al.,* 1997a).

7.3. Maturity Index

Physalis shows an intermediate behavior with increased respiration during ripening of climacteric fruits. The skin color of the cape gooseberry can be used as a maturity index (Galvis *et al.,* 2005). Harvesting should begin when the calyx begins to turn yellow, avoiding overmaturity (Bernal, 1991). With the fruit development fruit size, weight, soluble solids, acidity increases linearly up to full maturity stage. Visual harvest determination utilizes the synchronous color changing of both calyx and fruit, having nearly the same color. It should be done carefully to avoid the stem breaking and knocking off ripe fruits. Castañeda and Paredes (2003) observed fruit and color development in Granada (Cundinamarca state, Colombia). Fruits presented an intense green color during the first 35 days after anthesis, starting to change slowly to yellow, which was an intense yellow-orange in the skin and pulp at 64 days when consumer maturity was reached. At 84 days, fruit coloring was red-orange indicating the overmaturity stage. However, Singh *et al.,* 2012 reported the cape gooseberry should be harvested 6-8 weeks after anthesis when the fruit are well formed and substantially filled the calyx. Novoa *et al.* (2006) reported that fruit harvested at maturity grade 4 (Yellowish green colour) and those at grade 5 (yellow) dried at 18° C best conserved sucrose concentration. Sometimes premature fruit-fall is observed, probably related to overmaturity, abruptly changing soil moisture and climate and varietal hormonal factors. Preblossom spray with Ethrel (500 ppm) enhanced fruit ripening by 10 days (Garg and Singh, 1976). Angulo (2005) defined physiological maturity of the cape gooseberry, ready for picking, when the calyx, mostly green colored, presents yellowish stripes. At this stage it is first possible to see the fruit shape inside the husk by holding it against direct sunlight. At physiological maturity (56 days after anthesis) Castañeda and Paredes (2003) measured 12.7°Brix, 3.52 pH and 1.215 g of citric acid per 100 g fruit fresh weight.

1   2   3   4   5

Physalis (Goldenberry; Cape gooseberry) and Stage of Maturity/Ripeness Color

	1	2	3	4	5
Weight, g	1.78	1.94	2.07	1.76	2.16
per cent SS	10.8	11.7	12.8	13.6	13.8
pH	3.99	4.23	4.62	4.95	5.05
per cent TA	1.06	0.78	0.50	0.34	0.32
SS/TA	10.2	15.0	25.6	40.0	43.1

Chapter 8

Harvesting and Yield

Harvest initiates [depending on site conditions (principally temperature)] between 4 and 7 months after transplanting (Galvis *et al.*, 2005). Cape gooseberry fruits have a relatively low perishability, allowing for greater flexibility in harvesting. The fruit is harvested when it falls to the ground, but not all fallen fruits may be in the same stage of maturity and must be held until they ripen. It may take some experience to tell when the calyx-enclosed fruits are fully ripe. Properly matured and prepared fruits will keep for several months. In Colombia, starting with the second production year, the fruit size decreases, thus plants give their maximum yield in the first season (National Research Council, 1989), while whole culture is possible for 2 or 3 years. In sites with higher temperatures, first harvest peak is highest and then peaks decrease in intensity, whereas in colder climates it is the opposite (Fischer, 1995). It is recommended that the harvest is done two to three times per week during a harvest peak, in the early morning hours, avoiding picking during the rain or fruits with wet husks (Fischer and Almanza, 1993). Fruits should always be picked with their husk and with a peduncle with a maximum length of 25 mm (Icontec, 2004). In ecotypes and varieties where the peduncle is strongly attached, picking has to be done with small scissors. Harvesting can be accomplished by allowing the fruit to fall on fabric or plastic placed under the plants. Collection is either done by hand picking, or by gathering up the plastic and pouring the fruit into containers. It is recommended that the scissors be disinfected between each plant in plantations with disease infestation. For packaging the harvest, Icontec (2004) recommend plastic packing boxes with 7.00 to 7.5 kg fruits, and a maximum height of 250 mm.

In rainy or dewy weather, the fruit is not picked until the plants are dry. Berries that are already wet need to be lightly dried in the sun. Hand collection is preferable if the fruit is to be sold on the fresh market, to avoid bruising. At the peak of the season, a worker can pick 90 kg a day, but at the beginning and end of the season, when the crop is light, only 18 kg can be pick up in a day.

Physalis peruviana plants typically are heavy fruit producers. A single plant can produce up to 1.5 kg of fruits. According to other sources, a single plant may commonly yield from 130 to 300 fruits. Seedlings set 1,800 to 2,150 to the acre (228-900/ha) yield approximately 1,500 kg of fruit per acre. The average yield from a plant covering 2.5 square meters was found to be 545 g. The fruits are usually dehusked before delivery to markets or processors. Manual workers can produce only 4.5 kg to 5.5 kg of husked fruit per hour. Therefore, a mechanical husker, 4 to 5 times more efficient, has been designed at the University of Hawaii. The fruits can be sold with the husk left on as many chefs use the husk for decorative purposes. In test plantings at Ames, Iowa, the cape gooseberry yield averaged 1.1 kg per plant; equal to approximately 9 tons per acre (20.2 MT/ha). In India (Singh *et al.,* 2011), yields of 7.5 to 10 tons per acre (17-22.5 MT/ha) have been reported. Yields of 20 tons per hectare are common in South America, 33 tons has been achieved.

Plant growth, production and quality of cape gooseberry fruits can be achieved by application of plant growth regulators. The optimum GA_3 concentration spray and the critical stage of application for increase in growth and fruit yield of cape gooseberry are 100 ppm and one week after transplanting the seedlings. Application of GA_3 can therefore be recommended to help in improvement of growth and development of plant, hence resulting in increased fruit yield. (Wayama *et al.,* 2006).

Chapter 9

Insect Pests and Diseases

9.1 Insect Pests and their Control

Mites

Various types of insect pests can attack the calyx and fruit. The heaviest attack on calyx in Colombia is caused by the mite *Aculops lycopersia* in the peduncle zone, and also on leaves, giving it a greyish-ashy shade. Dry, warm weather with fast drying of affected tissues favors its incidence (Benavides and Mora, 2005). Possible control is with chemicals containing the active ingredients abamectin and fenbutatin oxide (Angulo, 2003).

Leaf Borer

The leaf borer *Epitrix cucumeris* is another insect that causes great damage to the calyx and leaves, producing orifices of irregular size (Almanza and Fischer, 1993). It can be controlled by products based on the active substances deltamethrin or benfuracarb (Angulo, 2003).

Thrips

Adult and larval thrips (*Frankliniella* sp.) can damage apical leaves, flowers and husks. The affected tissues take on a whitish coloration, later turning silver and finally dark (Benavides and Mora, 2005). Against thrips, Angulo (2003) recommended a biological control with *Chrysoperla externa*.

Fruit Borers

Cape gooseberry fruit is infected by the borer *Heliothis* sp., which perforates the husk and consumes the pulp during any phase of its maturation (Benavides and Mora, 2005). This pest can diminish yields by more than 20 percent, and without the protection of the husk all fruits would be destroyed in a few days. Other fruit borers, with minor incidence such as *Copitarsia* sp. (Benavides and Mora, 2005) and *Lineodes* sp. (Angulo, 2003), were observed. Zapata *el al.* (2002) recommended biological control with products containing *Bacillus thuringiensis.* For heavy attacks of these fruit borers, products based with deltamethrin can be applied (Angulo, 2003). This pest can diminish yields by more than 20 percent, and without the protection of the husk all fruits would be destroyed in a few days. Other fruit borers, with minor incidence such as *Copitarsia* sp. (Benavides and Mora, 2005) and *Lineodes* sp. (Angulo, 2003), were observed.

9.2 Diseases

Grey Mold

One of the most important diseases in fruits and calyx, reported in Colombia, is *Phoma* sp. (Zapata *et al.,* 2002). This disorder initiates in the point where the fruit is inserted onto the peduncle with a black ring and finally develops a white mycelium on the fruit. Grey mold (*Botrytis* sp.) forms necrotic irregular spots developing a mycelium of grey color that can cover the whole fruit and calyx (Zapata *et al.,* 2005). When fruit are cracked, *Botrytis* can enter easily in the unprotected pulp causing a bitter flavour (Angulo, 2003).

Soft Rot

A typical post-harvest disease of cape gooseberry in the Maharashtra region of India described by Rao and Subramoniam (1976) is *Fusarium equiseti* (Corda) Sacc., entering at points of injuries and bruises on stored berries and subsequently inciting a soft rot. In the fruit market of Aligarh (India), Sharma and Khan (1978) isolated on cape gooseberries the fungi *Alternaria alternata* (Fr.) Keissler, *Cladosporium cladosporioides* (Fres.) de Vries and *Penicillium italicum* Wehmer.

Grey Spot

Other diseases that affect the leaves and calyx are grey spots (*Cercospora* sp., possibly *Cercospora physalidis* [Ellis, 1971]). Grease spots, the most important leaf disease of cape gooseberry, which impacts more during high humidity seasons (Blanco, 2000); and the bacteria _Xanthomonas_ sp., causing grease spots (Zapata _et al.,_ 2005). Disease control is based on good agricultural practices (healthy plant material, sufficient plant distance, sanitary pruning, etc.) and preventive sprays of fungicides.

9.3 Wilting and Drying of Calyx

The descendant drying of the calyx apex in fruits right before the harvest, as described by Blanco (2000) in Colombia, is caused by a _Cladosporium-Alternaria_ complex. Most of the fruits with completely dried husks are abscised. Against _Alternaria,_ Blanco (2000) recommended planting more resistant cultivars and applying fungicides at the moment when the secondary dissemination of the disease occurs.

Other, more seldom-observed diseases, mainly on leaves and sporadically on the husks, are _Ralstonia solanacearum,_ and a mosaic virus ('mosaico de uchuva') similar to the potato virus X (PVX) (Zapata _et al.,_ 2002). Powdery mildew can be found on leaves of cape gooseberry (CRFG (1997).

After 33 days' storage of cape gooseberry, Lizana and Espina (1991) found fungus of the genera _Cladosporium, Penicillium, Botrytis_ and _Alternaria_ on fruits kept at 7°C, whereas those at 0°C showed a significantly lower percentage of infection.

9.4 Disorders

9.4.1 Chilling Injury

Cape gooseberry fruits resist low storage temperatures and no chilling injuries are observed at temperatures as low as 1°C. Thus, 30-day storage at 1°C conserved fruit in good conditions (Garzón and Villareal, 2009). Alvarado _et al._ (2004) found no chilling injury at 1.5°C during 16 days.

9.4.2 Other Physiological Disorders

High fruit losses of cape gooseberry, up to 50 percent of all fruits not suitable for export, result from fruit cracking and splitting mainly caused by high water content in fruit, especially during rainy seasons or after the first rainfalls immediately following dry seasons (Fischer, 2005). Also, deficient fertilization with calcium and boron increased the percentage of cracked fruit from 5.5 to 13.0, when any of these two elements were eliminated in the nutrient solution (Cooman *et al.,* 2005). In cape gooseberry pot culture in greenhouses, irrigation with Mg-deficient nutrient solution resulted in 11.3 percent cracked fruits at harvest compared to 0.79 percent in the Ca-deficiency treatment (Garzón and Villareal, 2009). Also, high nitrogen content in soil (due to excessive mineral or organic fertilization) can increase cracking, up to 30 percent (Gordillo *et al.,* 2004). High relative humidity conditions in the air were observed to cause fruit cracking, both in field and in storage (Fischer, 2005) due to transpiration inhibiting effects. This effect is more pronounced as low night temperatures occur (Kaufmann, 1972). Also, cuticle formation on the fruit is possibly negatively affected causing a low protective cuticular capacity under high air humidity conditions (Opara *et al.,* 1997). To minimize cape gooseberry fruit cracking and splitting before harvest an optimum nutrient application with calcium, boron, potassium and magnesium is required, along with an avoidance of high soil humidity and excessive nitrogen applications. Also, products that contain carboxylic acids, complemented with calcium and boron, have shown to reduce this fruit disorder (Guerrero *et al.,* 2007). Zapata *et al.* (2002) mentioned that at the end of storage, fruits could present damage due to dehydration, ruptures of the calyx and fruit cracking. Cracked fruits during storage can occur due to abrupt changes in RH or temperature (Fischer, 2005), especially when the RH is high.

Chapter 10

Postharvest Management

10.1 Postharvest Physiology and Quality

10.1.1 Respiration and Ethylene Production

Cape gooseberry ripening is associated with a conspicuous climacteric rise in carbon dioxide (Novoa *et al.,* 2006) and ethylene production (Trinchero *et al.,* 1999). Respiration peak of intact fruits occurred under field conditions in Colombia (Silvania, Cundinamarca department, 2,100 m a.s.l.) at 64 days after anthesis (Castañeda and Paredes, 2003). Fruits without a calyx respire more than with an attached calyx (Villamizar *et al.,* 1993); also, supposedly, the intact calyx generates lower levels of ethylene.

Ethylene acts as a promoter of fruit softening as a consequence of cell-wall weakening caused by the activity of hydrolases (Fischer and Bennett, 1991). During ripening of 'Rashbhori' cape gooseberry fruits, Majumdar and Mazumdar (2002) found that water and oxalate-soluble pectic substances decreased while polygalacturonase activity increased: the latter was highly correlated with ethylene evolution, but pectin methylesterase activity was not clearly related to fruit ripening. With an Argentinean ecotype, polygalacturonase and α-glucosidase activity were hardly noticeable whereas pectin methylesterae and α- and β-galactosidase reach activity similar to that in the tomato fruit (Trinchero *et al.,* 1999).

10.1.2 Ripening, Quality Components and Indices

The ripening of Cape gooseberry is associated with a conspicuousclimacteric rise in CO_2 and ethylene production. Its respiration rate ethylene bio synthesis can be classified as extremely high. Ethylene yields between 7 and 24 μmol n^{-1} per gram in the ripe/over ripe stages thus compare favorably with production rates reported for tomato. As the fruit colour turns green (chlorophyll) to yellowish orange (carotenoids) and progressive softening occurs, several cell wall changes occur. The most noticeable change of cape gooseberry fruit during maturation and ripening is the change in skin color from green to yellow (Fischer and Martínez, 1999), in stages 2 and 3 (Figure 10.1) when physiological maturity occurs and synchronously the calyx changes from light green to yellow, which makes this characteristic adequate for use as a maturity index as discussed before. In cape gooseberry fruit, at the same time as the color change, the weight of the fruit increases, reaching a maximum around maturity stage 5 (Figure 10.2) (Fischer and Martínez, 1999). Total soluble solids (TSS), in green fruits with 9.3°Brix, peaked with between stage 3 and 4 and then fell to about 13.7 at the overripe stage (Figure 10.3). Total titratable acidity (TTA), due to its demand in respiration (Kays, 2004), decreases constantly from 39.5 mval/100 mL in green to 17.6 mval/100 mL in overripe fruits. Interestingly, β-carotene content, taking into

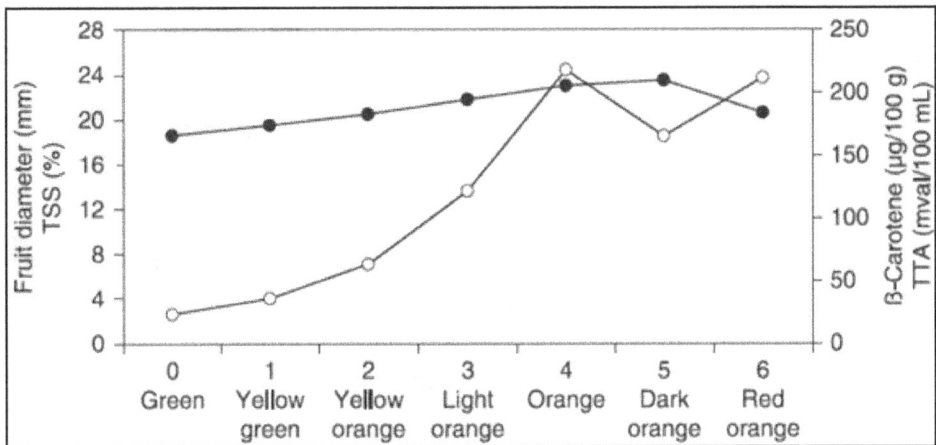

Figure 10.1: Changes in Diameter, Total Soluble Solids (TSS), Total Titratable Acidity (TTA) and β-carotene Content of Cape gooseberry Fruits during Six Ripening Stages.

Source: Data from Fischer, 1995

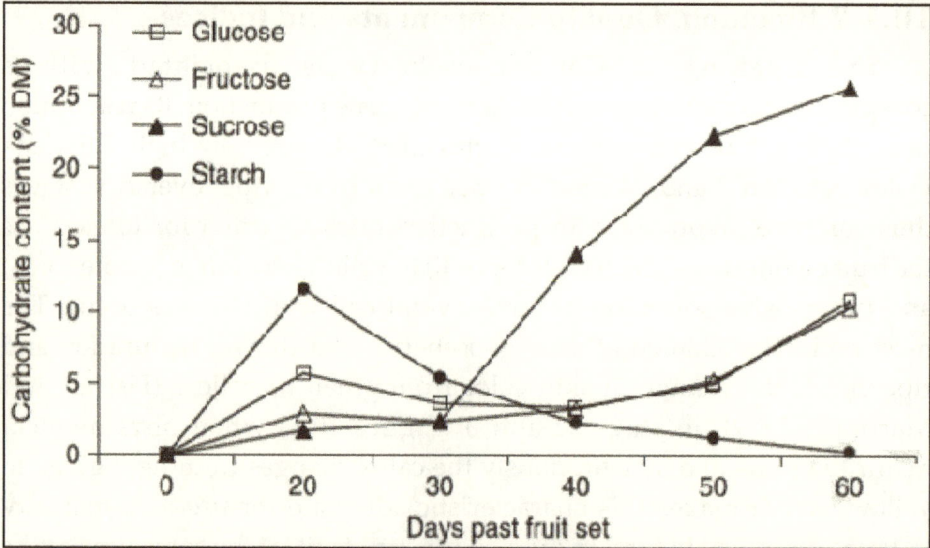

Figure 10.2: Developmental Changes of Carbohydrate Components of Cape gooseberry Fruits.
Source: Data from Fischer, 1995.

account that the cape gooseberry is classified as a carotenoid fruit (Fischer *et al.*, 1997b), peaks at the orange color stage, and after that its fall is interrupted because of a concentration effect on the decreasing fruit size at stage 6 (Figure 10.1).

Comparing carbohydrate patterns during fruit development, Fischer and Lüdders (1997) found an abrupt starch degradation from day 20 past fruit set, whereas sucrose content increased constantly, with high and increasing rates from day 30 (Figure 10.2). At harvest, at 56 days after anthesis, in Tunja (Colombia, Boyacá department, 2,690 m a.s.l.) fruits of the Colombia ecotype had 17.3°Brix, 2.0 percent TTA (8.6 SSC/ TAA), with a density of 1.06 g cm "3 (Almanza and Espinosa, 1995). Curiously, fruits of the Colombian-introduced African ecotypes Kenia and South Africa did not develop an intense yellow color (only dark ochre yellowish), ripened 10 to 15 days later and had lower SST and TTA content and density than the Colombian provenience (Almanza and Espinosa, 1995). Trillos *et al.* (2008) found in 46 cape gooseberry proveniences, kept in the Colombian Germplasm Bank of Corpoica, a medium TSS content of 12.2°Brix (maximum 16.9°, minimum 6.1°) and a juice content of 2.0 mL (maximum 4.0 mL, minimum 0.8 mL) per fruit. At harvest, Martínez *et al.*

(2008) observed a reduced total soluble solids content in fruits of plants deficient in boron fertilization, whereas those with a lack of phosphorous applications had higher total titratable acidity. A deficient phosphorous or calcium nutrition of plants in pot culture diminished the eating quality of cape gooseberry fruits more than a lack of magnesium in the nutrient solution (Garzón and Villareal, 2009). Pectinmethylesterase and α- and β-galactosidases reach activity level similar to tomato. Pectinmethylesterase and α-Arabinofuranosidase and β-glucosidase show lower activities but with increasing pattern during ripening (Gustavo *et al.,* 1999). Water, oxalate, acid and alkali soluble pectic substances as well as pectolytic enzymes activity and erhylene evolution were monitored in cape gooseberry throughout their development and ripening were studied by Majumdar and Majumdar (2012) and reported that water and oxalate soluble pectic substances were found to increase while those of acid and alkali soluble pectic substances decreased during ripening.

Simultaneously with the degradation of high molecular weight pectin, 5-6 fold increase in the polyglacturonase activity may not clearly related to fruit ripening. The increased level of polygalacturonase activity was highly correlated with ethylene evolution occurred prior to polygalactoronase synthesis in fruit tissue.

10.2 Handling and Grading

Generally, the berries are not washed or disinfected, regardless of whether the husk is attached or not (Gallo, 1992). A very high percentage of fruits are usually clean owing to protection from the calyx. Washing and disinfecting with low concentrations of chlorine (20 to 50 ppm) decrease the microbial contamination but eliminate the natural anti-feeding compounds of the fruit (see point 1.1), thus Alvarado *et al.* (2004) found increased fruit pulp dehydration and diminished preservation time after disinfection. In all fruits destined for export with the calyx, the husk is carefully opened at the apical end and checked for defects (physical, physiological, pathological and entomological), size and color (Galvis *et al.,* 2005). Just like the fruit, the husk has to be free of defects. Generally, in the course of this process, fruit with discolorations (too green or too orange) or cracks are sorted out. Fruits have to be firm and fresh in appearance, with a smooth and shiny skin. If the calyx is present, the peduncle must not exceed 25 mm in length (Icontec, 1999).

In accordance with import requirements and/or the distance to the import country, the color of the calyx can be light greenish, normally straw-colored. The size is determined by the maximum diameter of the equatorial section of the fruit. For size code diameters, Icontec (1999) determined for caliber A d" 15.0 mm, B 15.1 to 18.0 mm, C 18.1 to 20.0 mm, D 20.1 to 22.0 mm, and for caliber E ≥22.1. Unit fruit weight is not standardized, but fruits (without the calyx) of 6 g are preferred; however in periods of high demand, fruits sometimes only weigh 4.5 g.

10.3 Calyx Drying

Since cape gooseberry is mainly exported with an attached calyx, conserving the quality characteristics and protecting the fruit from damage, calyx drying is of great importance in the postharvest life of this fruit. This is the reason that cape gooseberry importers require fruits with totally dried husks. The calyx, which is not edible, has a very similar carbohydrate pattern to that of the fruit (Fischer and Lüdders, 1997), and can weigh up to 2 g, although dried ones have only one-tenth this weight.

Generally, calyx drying is done after fruit grading, before or after packaging (commonly done in different types of small plastic baskets). It is recommended that drying be done after packing due to the ease of packing fruit with a fresh calyx compared to fruit with dry husks (Galvis *et al.*, 2005). Colombian fruit export firms, generally, dry calices by convection, with forced air with RH lower than 50 percent, but the temperature can vary. Good results were found by Novoa *et al.* (2002) drying fruit calices at an air temperature of 24°C for 6 hours without affecting physico-chemical and sensorial characteristics during or after posterior storage. Longer drying periods cause fruit weight losses and accelerate the ripening process (Galvis *et al.*, 2005). Ten hours of drying with temperatures up to 25°C can be used for fruits with green or wet calices. In the drying system by convection, a chamber (or tunnel) is used into which hot air is injected on one end through fans. The air, which passes through the small baskets packed with fruits, is removed with extractors at the other side. Novoa *et al.* (2002) observed a higher incidence of *Botrytis* on fruits that were dried at 18°C compared to 24°C. For the local market, calyx drying can be done on tables or on wire nettings. With an air temperature of 12°C, the aeration can last 8 hours using fans and up to 3 days without artificial air movement (Herrera, 2000).

10.4 Packaging

The best packaging for Cape gooseberry fruit is the natural attached dried calyx: it protects the fruit not only from physical damage but also against fungus, and favors the modification of the atmosphere surrounding the fruit (Galvis *et al.,* 2005). For distribution to the wholesale markets, cape gooseberries are preferentially packed in plastic boxes or also those made of cardboard or wood (for 8 to 10 kg fruit weight). Icontec (2004) recommended a maximum height of 25 cm, with a capacity of 8 kg for fruit with a calyx and 10 kg without a calyx. For consumers, fruits are packed in perforated small plastic fruit containers or in small plastic baskets with a total maximum fruit weight of 200 g (with calyx) and 500 g (without calyx). For export, packaging specifications depend on the importer and the target country requirements, normally small plastic baskets with 125 to 250 g fruit weight of cape gooseberries with calyx are used. Fruit packed in these baskets are commonly covered with a plastic film (polyvinyl chloride or micro-perforated polypropylene) (Herrera, 2000). The baskets are grouped into corrugated fiberboard boxes (*e.g.* eight baskets with 125 g each results in 1 kg fruit weight). For transport handling, these boxes are packed into mini-containers that are sized in accordance with the dimensions of the pallets. Requirements for package material (type, hygiene, etc.) and labeling must be in accordance with international standards, especially with those of the import country.

10.5 Storage

The fruits of *Physalis peruviana* are long-lasting. The fresh fruits can be stored in a scalded container and kept in a dry atmosphere for several months without refrigeration. They will still be in good condition. If the fresh fruits are to be shipped, it is best to leave the husk on for protection. The unhusked fresh fruits of *Physalis ixocarpa* can be stored in single layers in a cool dry atmosphere for several months. Mexican and Central American people may pull up the entire plant with fruits attached and hang it upside-down in a dry place until the fruits are needed.

Avoid buying fruits in green color which are not ripe. While preparing of fruit for use peel back the parchment like husk and rinse. At the time of purchasing its outer covering should not be torn and must be in good condition also it should keep its shape. The fruit can be kept more than 6 months in a ventilated spot by prolonging the ripening when it is being protected by its calyx.

10.5.1 Temperate and Relative Humidity

Optimum conditions for medium- and long-term storage of cape gooseberry fruit (<6 months) vary between 2°C and 4°C, and 80 to 90 percent RH (Galvis *et al.,* 2005). In fruits stored with calyx, a high RH can cause diseases, which highlights the importance of good calyx drying before storage. A RH of 68 percent and 88 percent had no significant effect on fruit quality during 16-day-storage of Cape gooseberries at 1.5°C (Alvarado *et al.,* 2004).

Mercantila Publishers (1989) recommended an ideal storage or transport temperature of 12 to 15°C for 1 to 2 months, and for 30-day storage in Colombia temperatures between 4°C and 6°C showed good results. The transport temperature for Colombian cape gooseberries is commonly set at ranges of 8 to 12°C.

In general, fruits with a calyx can be stored for a longer duration at different temperatures, 6°C (Villamizar *et al.,* 1993), 12°C (Novoa *et al.,* 2006) and 19°C (García and Torres, 2005, cited by Galvis *et al.,* 2005). Villamizar *et al.* (1993) reported that fruit held at 6°C could be stored for 30 days with an attached calyx, but only for 20 days without a calyx, both at 18°C and 70 percent RH.

During a 33-day-storage period at 0 and 7°C, Cape gooseberry fruit titratable acidity content decreased more at the higher temperature regime where a higher respiratory rate was also measured, supposedly responsible for the higher degradation of the organic acids used as a respiratory substrate (Kays, 2002). Also, the higher fungus infection in this treatment (7°C) could be due, in part, to the lower acidity of these fruits (Osterloh *et al.,* 1996).

Owing to its resistance to chilling injury, at as low as 1 to 2°C (Alvarado *et al.,* 2004), Cape gooseberry fruits tolerate the quarantine cold treatment T107a well, which is required by APHIS-USDA as a condition to export these fruits from Colombia to USA. The treatment involves maintaining fruit pulp temperature at <2.22°C during definite periods (*e.g.* 14 days at 1.10°C, 16 days at 1.67°C or 18 days at 2.20°C), with the purpose of killing the Mediterranean fruit fly larva (Flórez, 2005). The treatment must be implemented in containers that are automatically cooled or in refrigerated ships during transit (USDA, 2004). In some cases the quarantine cold treatment is applied after air transport in Atlanta, Georgia (USA). The cold treatment for Cape gooseberry fruit exported to the USA possesses variants, *e.g.* fumigation + cold treatment (T108), cold treatment + fruit fumigation (T109) and quick freezing (T110) (USDA, 2004). Also irradiation (T105-b-4) treatment is allowed to kill fruit flies in the cape gooseberry before entering the USA ports (Flórez, 2005).

10.5.2 Controled and Modified Atmosphere

There is little information on the use of modified and controlled atmosphere storage of cape gooseberry. Gallo (1992) recommended an atmosphere of 3 to 10 percent CO_2 and 3 percent O_2 at 5 to 8°C and 85 to 90 percent RH. During a 4-week-storage in modified atmosphere at 7°C using three kinds of plastic film (polyethylene terephthalate-polyethylene, bi-oriented polypropylene-polyethylene, and polyolefin) and four atmosphere mixtures (5 percent CO_2 and 5 percent O_2; 5 percent CO_2 and 10 percent O_2; commercial mixture and ambient air), the best results were shown with calyx-attached fruits (higher firmness) packed in polyolefin film (lower water loss), with no differences among active modified atmospheres (Lanchero *et al.*, 2007). Also, cape gooseberries packed in plastic baskets wrapped with microperforated polypropylene film lost only 3.6 percent of fresh weight during a 33 days' storage at 7.5°C, whereas fruit without film wrapping lost 5.0 percent after 16 days (López and Páez, 2002).

10.5.3 Ethylene

The Cape gooseberry was classified by Gallo (1992) as a high producer of ethylene, up to 100 $\mu L\ kg^{-1}\ h^{-1}$. This has to be taken into account when using modified atmosphere packaging for advanced ripened fruits because it can result in the accumulation of ethylene, thus counteracting the beneficial effects of ripening delay. Application of the ethylene antagonist 1-methylcyclopropene (1-MCP) gas delayed the onset of ethylene climacteric in mature green fruits and transiently decreased ethylene production in yellow and orange cape gooseberry fruits (Gutierrez *et al.*, 2008). Also, this study revealed that 1-MCP application did not prevent decay in orange-colored fruits but reduced its incidence, thus suggesting that it may influence pathogen infection and development in ripe cape gooseberries. Because of its high ethylene production, it is not recommended that cape gooseberry be mixed with other fruits that are sensitive to exogenous ethylene, *e.g.* the purple passion fruit (Shiomi *et al.*, 1996), since it accelerates endogenous ethylene production and can thus accelerate their ripening. If for some reason they must be mixed, ethylene scrubbers may reduce damage (Thompson, 2002).

10.5.4 Waxing

Waxed fruit, without the husk and using bee honey, which was emulsified with vegetable oil (with a ratio of 1:8) beforehand, better maintained the sensorial characteristics and the original fresh weight in storage, compared to non-waxed fruit (Rodríguez, 2003). Waxed fruits maintained total soluble solids and firmness for 41 days.

10.6 Processing

The Cape gooseberry fruit which is acid sweet in taste and with a pleasant flavor is typically consumed fresh, whole or dried, without the calyx, but with the skin. Fruits of *Physalis peruviana* are juicy, widely astringent and sweet with a pleasant blend of acid. The overall quality of the fruit is good. In addition to being canned whole or preserve it d can be used in order to prepare desserts (pies, cakes, jellies, jams, puddings, chutneys, sauces, ice cream, yoghurt, (Annual report of CITH, Srinagar, 2010-11) (Camacho, 2000). Its flavor has been defined as a pleasant, unique tomato/pineapple like blend. The osmo dehydrated products fruit can be used as a raisin and yeast substantiate, though it is not so sweet. These fruits are excellent when differ in character, drier and eaten. British use the husk as a handle for dipping the fruit in icing. The fruits taste excellent when dipped in chocolates.

Ripe good-quality fruits can be used as a dried food, similar to raisins and exported as such. Vitamin E-impregnated cape gooseberry fruits (using an isotonic dissolution of sucrose in the aqueous phase of the emulsion), were juicier, sweeter and less acidic than natural samples (Restrepo *et al.*, 2008). Due to their 'exotic touch', it is common to use Cape gooseberry fruits, half opened or without the husk, for decoration purposes on cold buffets in restaurants and hotels.

Modern consumers are interested in healthy food to ensure healthy ageing. Cape gooseberry fruits fortified with vitamin E by vaccum impregnation (IV). Vaccum packaging with significant colour (dark) and texture (soft) (Restrepo *et al.*, 2009.)

10.7 Novel Value Added Products

1. Cape gooseberry Jam

Ingredients

- ☆ Cape gooseberries as required
- ☆ Orange, Pineapple (if desired)
- ☆ Sugar as required.
- ☆ Water as required.

Method

1. Take equal weights of all fruits.
2. Wash all the fruit and bring it to boil.
3. Add 500 g sugar and 250 ml water per kilogram of fruits added.
4. Reduce the heat and let simmer for a few minutes.

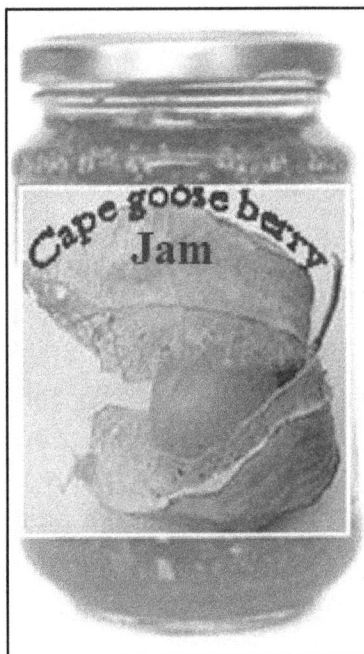

5. Place the fruit into the syrup, return to the boil and allow it to cook for more than 25 minutes.

6. Check the thickness of the jam, if it's ok, then remove it from heat and seal it in an air tight container.

2. Cape gooseberry Cake

Ingredients

40 large Cape gooseberries or 2 cups of Cape gooseberry (sliced).

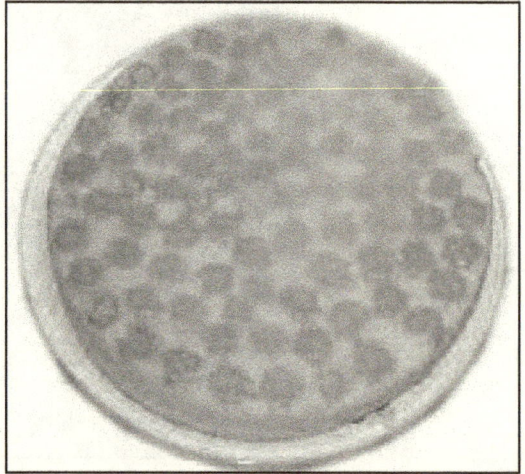

☆ 1/2 cup softened butter

☆ 2 tablespoons jaggery bits as required

☆ 1/2 cup powdered sugar

☆ 1½ cups of flour (maida)

☆ 1½ tsp of baking powder

☆ 1/2 cup semolina (sooji)

☆ Pinch of salt

☆ 3 eggs

☆ 1½ tablespoons of milk

☆ 3-4 drops of vanilla essence

☆ 2 tablespoons whipping cream if needed

Method

1. With a paring knife carefully tear each gooseberry across till the stem but keep both halves attached.

2. In a pan simmer the gooseberries over a very low fire till some of the juice is released and the berries become soft still keeping their shape and not pulpy.

3. Take it off from the fire, pour off the juice and keep it aside.

4. Melt some butter and pour it into the cake tin.

5. Arrange the gooseberries on the base of the tin and sprinkle the jaggery bits on the top.

6. Heat the oven to 350° C for 10 minutes.

7. Cream the rest of the butter and sugar together till it becomes light and fluffy.

8. In a bowl mix the flour and baking powder.

9. Add the sooji and a pinch of salt.

10. Beat the two eggs and add 1½ tablespoons of the juice obtained from the cooked gooseberries and 1½ tablespoons of milk to the mixture.

11. Add the vanilla essence as well.

12. Now mix the flour mixture into the butter mixture in three parts alternating with the egg mixture.

13. When well mixed spoon the mixture over the gooseberries in a cake tin.

14. Place in the oven and cook till done, about 30 minutes.

15. Let it cool and turn out on to a serving plate.

16. Top with whipping cream.

17. Decorate it with some nutritious nuts and serve.

3. Cape gooseberry Chutney

Ingredients

- ☆ 1 cup of Cape goose-berries
- ☆ 25ml of vegetable oil
- ☆ 1/4 tsp mustard seeds
- ☆ 1/4 tsp cumin seeds
- ☆ Chopped onion as required
- ☆ 1 tsp chopped ginger
- ☆ Chopped green chili as required
- ☆ Salt as required
- ☆ 1/4 tsp red chili powder

☆ 1/4 tsp cumin powder

· ☆ 1/4 tsp coriander powder

Method

1. Take a pan, heat around two tbsps of oil.

2. Once oil gets heated add the mustard and cumin seeds to it.

3. After the mustard splutters, add the chopped onion to the pan.

4. Cook until it becomes transparent, then add the chopped chilli and ginger to the mixture.

5. Cook for 2-3 minutes, add the powdered spices and salt.

6. Cook in a low flame, then add the sliced gooseberries and sugar (optional).

7. Gently saute over a low flame for a few minutes.

4. Cape gooseberry and Apple Crumble

Ingredients

☆ 1 kg of dessert apples

☆ 4-5 tbsp dried strawberries

☆ 4 tbsp of orange juice

☆ 1 cup of cape gooseberries

☆ Sugar as required

☆ 100 g Plain flour

☆ 100 g unsalted butter, diced

☆ 55 g of caster sugar

☆ 100 g rolled oats

☆ 50 g or walnuts or cashew or almonds

☆ Grated zest of 1 orange

☆ Custard

Method

1. Preheat the oven to 200°C.

2. Peel the apples, cut them into thick wedges and remove the core.

3. Place them in a saucepan with the strawberries and orange juice.

4. Cover the pan and cook over a low heat for till the apples start to soften and release their juices.

5. Add the whole cape gooseberries and sugar to the apple mixture and stir.

6. Transfer the fruit filling to a baking dish.

7. In a mixing bowl, rub the flour and butter together until the mixture resembles fine breadcrumbs.

8. Stir in the sugar, oats, walnuts and orange zest.

9. Then mix in about 1 tbsp water to give a very rough, crumbly mixture.

10. Spread the crumble topping gently and evenly over the fruit without pressing it.

11. Bake for 20–25 minutes till the topping becomes golden brown and the fruit juice is bubbling up round the edges.

12. Serve hot, with custard.

5. Cape gooseberry Sauce

Ingredients

☆ 1 kg Cape gooseberry

☆ 1 ½ cups brown sugar

☆ Chopped onions 4

☆ 1 tsp ground ginger

☆ Salt as required

☆ Cayenne pepper as required.

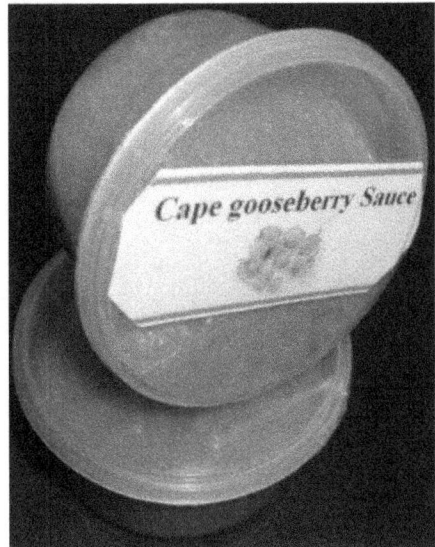

Cape gooseberry Sauce

Method

1. Wash and drain peeled gooseberry.

2. Take a saucepan, melt some butter in it.

3. Add gooseberry, onions and sugar into the saucepan.

4. Cook until the berries get soften, continue to simmer by stirring occasionally until sauce thickens slightly.

5. If needed you can strain the sauce.

6. Osmo Dehydrated Cape gooseberry

Osmo dehydrated products of Cape goosberry having excellent crisp, sweet and tangy taste with very good flavour can be prepared from well matured and ripe but firm fruits. Processing technology for making osmo dehydrated product is given in flow diagram.

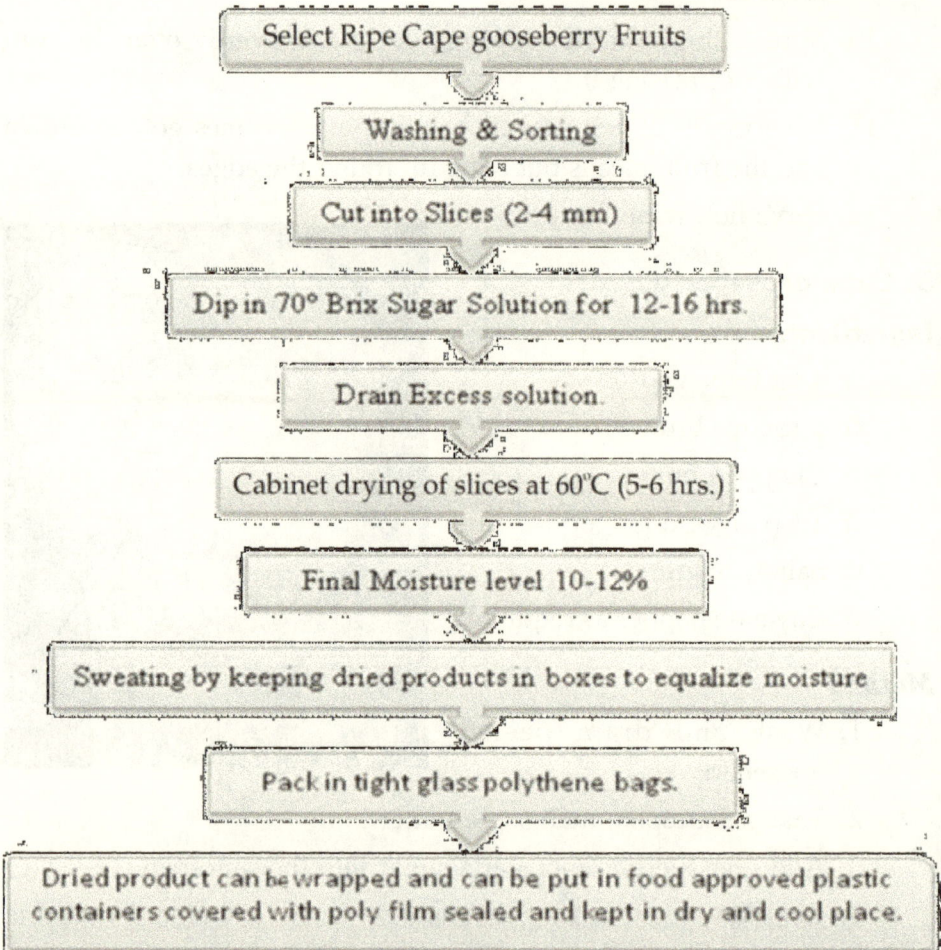

Select Ripe Cape gooseberry Fruits

↓

Washing & Sorting

↓

Cut into Slices (2-4 mm)

↓

Dip in 70° Brix Sugar Solution for 12-16 hrs.

↓

Drain Excess solution.

↓

Cabinet drying of slices at 60°C (5-6 hrs.)

↓

Final Moisture level 10-12%

↓

Sweating by keeping dried products in boxes to equalize moisture

↓

Pack in tight glass polythene bags.

↓

Dried product can be wrapped and can be put in food approved plastic containers covered with poly film sealed and kept in dry and cool place.

Flow Diagram for Preparation of Osmo Dehydrated Cape gooseberry

7. Cape gooseberry Fruit Bar

"Cape gooseberry Fruit Bar" has appealing texture, flavour, taste, aroma and has very good acceptable physical and chemical characters with excellent mouth refreshing ability. The final product does not stick to the teeth and gums when eaten. For preparing best quality Cape gooseberry fruit bar, fruits should have ideal colour, texture and flavour. Fruit should be ripe but not over ripe. Husk of the fruits is removed before handling of fruits for processing. Selected fruits are washed and treated with 100 ppm sodium hypochlorite for one minute. Bruised portion of fruits are discarded. For retention of colour and avoiding browning of product cape gooseberry halves are dipped in ascorbic acid (5 per cent Wv) and citric acid (5 Wv per cent) for 30 seconds. The stuff is then steam blanched for 5 minutes. Puree the fruit halves in a blender or processor until smooth slurry is produced. The puree/slurry is drained and passed through screen pulper. The final puree is concentrated with sugar up to 65±2°Brix. The final concentrate is spread in food approved plastic or steel trays about 4 mm thickness in tunnel dryer (sun drying) for 22-24 sun shine hours (temperature 48-50°C and relative humidity of 30 per cent). Flow diagram describes the processing of make fruit bar of Cape gooseberry.

Selection of fruits (Ripe, rich in colour, texture and flavor

Remove husk, washed and treated with 100 ppm sodium hypo chlorite for 1 minute

Remove stems and bruised portions

Halves are dipped in Ascorbic acid (5% Wv) and citric acid (5% Wv) for 30 seconds

Steam blanched for 5 minutes

Puree the fruit halves in blender or processor until smooth slurry produced

Drained and passed through screen pulper

Puree concentrated with sugar up to 55° Brix ±2

Concentrate spread in trays about 4 mm thick (plastic or stainless steel) in tunnel dryer (Sun drying) for 22-24 sun shine hours (Temperature 48-50° C and RH 30%) or cabinet dryer at 60° C for 5-6 hours (in stainless steel trays).

Final moisture 14-18 %

Dehydrated product (fruit bars) is wrapped and can be packed in 25µ polyethylene film or in food approved plastic containers covered with poly film, sealed and kept in dry and cool place.

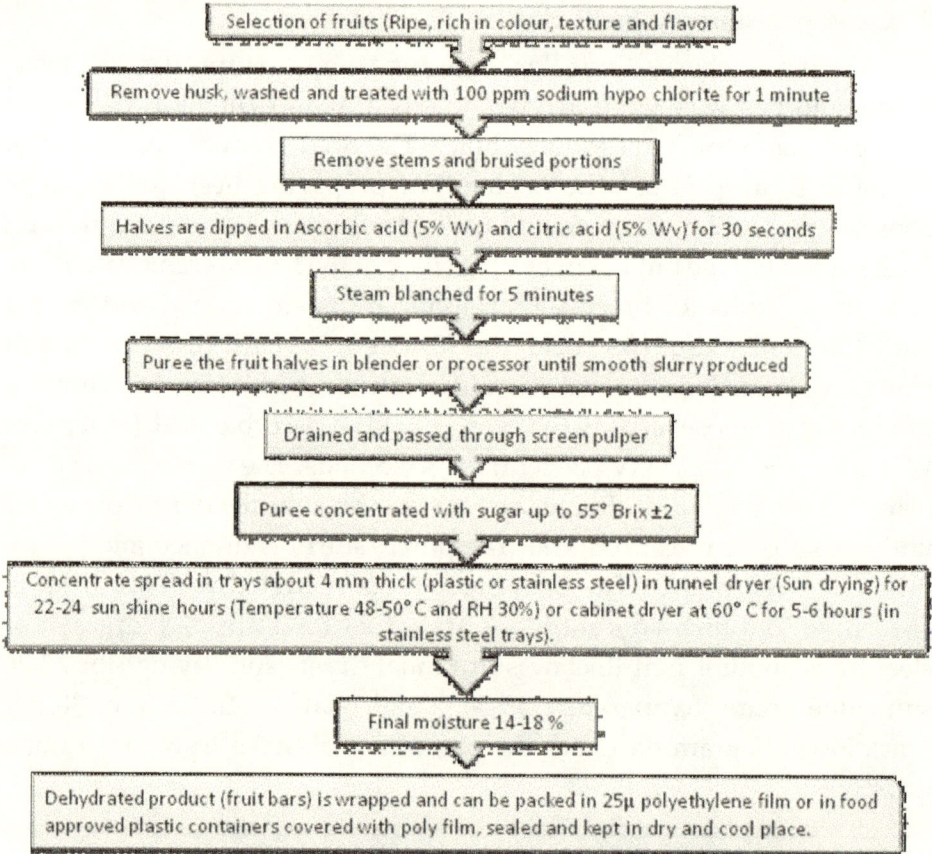

Flow Diagram for Preparation of Cape gooseberry Fruit Bar

Chapter 11

Future Scope

Growing consumers demand for "New" and unique fruits and vegetables is spurring a need for increased information on cultural techniques for these crops. The increasing demand for new and alternate crops, together with environmental concerns particularly that of climate change, makes the evaluation of sustainable agriculture production techniques desirable.

Systematic approaches to introduction and adaptation of new crops contributes to an increase in the diversity of agricultural systems and to offering new alternatives to farmers and markets with crops that may have a high value and for which there is no over production. Therefore new crops can result in increase in income to farmers, contribute to a more environmentally friendly agriculture, reduce the risk of crop failure and increase the botanical knowledge. There are many new crops for tropical and sub tropical regions that can present desirable attributes to be introduced as new crops in temperate climates. Among these, cape gooseberry, grown for edible fruits is usually short cycled annual crop, which in its region of origin is adapted to a wide altitude range including erolytic warm areas with an intense solar radiation to humid and cloudy environment. Its wide range of versatile adaptation and use for table purpose and processing form and increasing demand in exotic fruit market can gain good prospects for expansion of cape gooseberry as a new cash crop in temperate regions.

To promote its production and marketing, the following future need and challenge include:

1. Seletion of varaties/cultivars bearing large size fruits having sweet and pleasant flaavour without cracking.

2. Storage facilities which can be protecting its shelf life and availability.

3. Further study/research on vitamins, antioxidants and mineral content of the fruit in order to better promote its nuteraceutical benefits.

4. Standadization making novel value added products.

Literature Cited

Ahmad S., Malik A., Yasmin R., Ullah N., Gul W., Khan P. M., Nawaz H. R. and Afza N. (1999), 'Withanolides from *Physalis peruviana* ', *Phytochemistry,* 50, 647 – 651.

Almanza P. and Espinosa C. J. (1995), 'Desarrollo morfológico y análisis fisicoquímico de frutos de uchuva (*Physalis peruviana* L.) para identificar el momento óptimo de cosecha', Thesis, Facultad de Agronomía, Universidad Pedagógica y Tecnológica de Colombia, Tunja.

Almanza P. and Fischer G. (1993), 'Nuevas tecnologías en el cultivo de la uchuva *Physalis peruviana* L.' *Agro-Desarrollo,* 4 (1–2), 292 – 304.

Alvarado P. A., Berdugo C. A. and Fischer G. (2004), 'Efecto de un tratamiento de frío (a 1,5°C) y la humedad relativa sobre las características físico-químicas de frutos deuchuva *Physalis peruviana* L. durante el posterior transporte y almacenamiento', *Agron Colomb,* 22, 147 – 159.

Angulo R. (2003), *Frutales exóticos de clima frío,* Bogotá, Bayer CropScience, 27 – 48.

Angulo R. (ed.) (2005), *Uchuva – El cultivo,* Bogotá, Universidad de Bogotá Jorge Tadeo Lozano.

Anonymous 2011-12. Annual Report of Central Institute of Temperate Horticulture. Srinagar, J and K.

Arun M. and Asha V. V. (2007), 'Preliminary studies on antihepatotoxic effects of *Physalis peruviana* Linn. (Solanaceae) against carbon tetrachloride induced acute liver injury in rats', *J Ethnopharmacol*, 111, 110 – 114.

Ayala, C. (1990), Evaluation of Three Planting Distances and Three Systems of Pruning in Cape gooseberry Under Greenhouse Conditions. *ISHS Acta Horticulture* 310.

Bartholomäus A., de la Rosa A., Santos J. O., Acero L. E. and Moosbrugger W. (1990), *E l manto de la tierra – Flora de los Andes. Guía de 150 especies de la flora andina*, Bogotá, Edn Lerner.

Basra, A.S. ed. (1999). *Heterosis and Hybrid Seed Production in Agromatic Crops*. Binghamton, NY: Food Products Press.

Baumann T. W. and Meier C. M. (1993), 'Chemical defence by withanolides during fruit development', in *Physalis peruviana* ', *Phytochemistry*, 33 (2), 317 – 321.

Benavides M. A. and Mora H. R. (2005), 'Los insectos-plaga limitantes en el cultivo de la uchuva y su manejo', in Fischer G., Miranda D., Piedrahíta W. and Romero J., *Avances en cultivo, poscosecha y exportación de la uchuva (Physalis peruviana L.) en Colombia*, Bogotá, Unibiblos, Universidad Nacional de Colombia, 83 – 96.

Bernal J. A. (1991), 'Agronomic aspects of the cultivation of the uchuva, *Physalis peruviana*, on the high plateau of the Colombian departments of Cundinamarca and Boyacá', in Hawkes J G, Lester R N, Nee M and Estrada N, *S olanaceae III: Taxonomy, Chemistry, Evolution*, Royal Botanic Gardens Kew and Linnean Society of London, 459 – 460.

Blanco J. (2000), 'Manejo de enfermedades', in Flórez, V J, Fischer G and Sora A D, *Producción, poscosecha y exportación de la uchuva (Physalis peruviana L.)*, Bogotá, Unibiblos, Universidad Nacional de Colombia, 57 – 65.

Blum, A. (1988), *Plant Breeding for Stress Environments*. Boca Raton, FL: CRC Press.

Branzati, E.C. and A. Manaresi. (1980). L'alchechengi. *Frutticoltura* 42 (3-4):59.

Brondo-Leiva, M., J. Prohens, and F. Nuez. (2001), Genetic Analyses Indicate Superiority of Performance of Cape gooseberry (*Physalis peruviana* L.) Hybrid. *Journal of New Seeds,* Vol. 3(3).

Brücher H. (1989), *Useful Plants of Neotropical Origin and their World Relatives,* Berlin, Springer. 275 – 277. Camacho G. (2000), 'Procesamiento' in Flórez, V. J., Fischer G. and Sora A. D., *Producción, poscosecha y exportación de la uchuva (Physalis peruviana* L.), Bogotá, Unibiblos, Universidad Nacional de Colombia, 129 – 145.

Camacho G. (2000), 'Procesamiento' in Flórez, V. J., Fischer G. and Sora A. D., Producción, poscosecha y exportación de la uchuva (*Physalis peruviana* L.), Bogotá, Unibiblos, Universidad Nacional de Colombia, 129 – 145.

Camacho G. and Sanabria G. (2005), 'Alternativas de procesamiento y transformación de la uchuva', in Fischer G., Miranda D., Piedrahíta W. and Romero J., *Avances en cultivo, poscosecha y exportación de la uchuva (Physalis peruviana* L.) *en Colombia,* Bogotá, Unibiblos, Universidad Nacional de Colombia, 191 – 203.

Carman E. (1980/1981), 'Poha jam *Physalis peruviana,* exotic plants', *Pacific Hort,* 41 (4), 9 – 10. Castañeda G. E. and Paredes R. I. (2003), Estudio del proceso respiratorio, principales ácidos orgánicos, azúcares y algunos cambios físico-químicos en el desarrollo del fruto de uchuva (*Physalis peruviana* L.), Thesis, Bogotá, Facultad de Agronomía, Universidad Nacional de Colombia.

CCI (2000), 'El mercado de la uchuva', Exótica, July–September, 6 – 9.

CCI (2002), 'Uchuva – Perfil del producto', Inteligencia de Mercados, 13, 1 – 12.

Chattopadhyay, T.K. (1996). Cape gooseberry. In: *A Textbook on Pomology.* Vol. II, ed. T.K. Chattopadhyay, Calcutta: Kalyani Publishers, pp. 309-314.

Cooman A., Torres C. and Fischer G. (2005), 'Determinación de las causas del rajado del nfruto de uchuva (*Physalis peruviana* L.) bajo cubierta. II Efecto de la oferta de calcio, nboro y cobre', *Agron Colomb,* 23 (1), 74 – 82.

CRFG (1997) 'Cape gooseberry *Physalis peruviana* L.' California Rare Fruit Growers, 1 – 3.

Durán F. (ed.) (2007), ' Manual curativo con frutas y plantas medicinales ', Bogotá, Grupo Latino Ltda. 212 – 213.

Ellis M. B. (1971), Dematiaceous Hyphomycetes, Kew, Surrey, Commonwealth Mycological Institute.

Everett T. H. (1981), The Botanical Garden Illustrated Encyclopedia of Horticulture, Vol. 8. New York, Garland Publishing.

FAO (1982), ' *Physalis peruviana* L.', in Fruit-Bearing Forest Trees: Technical Notes, Rome, FAO Forestry Paper, 34, 140 – 143.

Fischer G. (1995) 'Effect of root zone temperature and tropical altitude on the growth, development and fruit quality of cape gooseberry (*Physalis peruviana* L.)' Ph.D. thesis, Berlin, Humboldt University.

Fischer G. (2000), 'Crecimiento y desarrollo', in Flórez, V. J., Fischer G. and Sora A. D., Producción, poscosecha y exportación de la uchuva (*Physalis peruviana* L.), Bogotá, Unibiblos, Universidad Nacional de Colombia, 9 – 26.

Fischer G. (2005), 'El problema del rajado del fruto de la uchuva y su posible control', in Fischer G., Miranda D., Piedrahíta W. and Romero J., Avances en cultivo, poscosecha y exportación de la uchuva (*Physalis peruviana* L.) en Colombia, Bogotá, Unibiblos, Universidad Nacional de Colombia, 55 – 82.

Fischer G., Ebert G. and Lüdders P. (2000), 'Provitamin A carotenoids, organic acids and ascorbic acid content of cape gooseberry (*Physalis peruviana* L.) ecotypes grown at two tropical altitudes', *Acta Hort,* 531, 263 – 267.

Fischer G., Ebert G. and Lüdders P. (2007), 'Production, seeds and carbohydrate contents of cape gooseberry (*Physalis peruviana* L.) fruits grown at two contrasting Colombian altitudes', *J Appl Bot Food Qual,* 81 (1), 29 – 35.

Fischer G., Lüdders P. and Gallo F. (1997b), 'Qualitätsveränderungen der Kapstachelbeere während der Fruchtreifung', Erwerbsobstbau, 39 (5), 153 – 156.

Fischer G., Lüdders P. and Torres F. (1997a), 'Influencia de la separación del cáliz de la uchuva (*Physalis peruviana* L.) sobre el desarrollo del fruto', Revista Comalfi, 24(1–2), 3– 16.

Fischer G. and Almanza P. J. (1993), 'La uchuva (*Physalis peruviana* L.) una alternative promisoria para las zonas altas de Colombia', *Agricultura Tropical,* (30) 1, 79 – 87.

Fischer G. and Lüdders P. (1997), 'Developmental changes of carbohydrates in cape gooseberry (*Physalis peruviana* L.) fruits in relation to the calyx and the leaves', Agron Colomb, 14, 95 – 107.

Fischer G. and Martínez O. (1999), 'Calidad y madurez de la uchuva (*Physalis peruviana* L.) en relación con la coloración del fruto', Agron Colomb, 16, 35 – 39.

Fischer R. L. and Bennett C. A. (1991), 'Role of cell wall hydrolases in fruit ripening', *Annu Rev Plant Physiol Mol Biol,* 42, 675 – 703.

Flórez E. (2005), 'Regulación fitosanitaria para la uchuva (cape gooseberry) con destinoal mercado de los Estados Unidos de América', in Fischer G., Miranda D., Piedrahíta W. and Romero J., Avances en cultivo, poscosecha y exportación de la uchuva (*Physalis peruviana* L.) en Colombia, Bogotá, Unibiblos, Universidad Nacional de Colombia, 205 – 210.

Fouche G., Cragg G. M., Pillay P., Kolesnikova N., Maharaj V. J. and Senabe J. (2008), ' *In vitro* anticancer screening of South African plants', *J Ethnopharmacology,* 119, 455– 461.

Franco L. A., Matiz G. E., Calle J., Pinzón R. and Ospina L. F. (2007), 'Actividad antin famatoria de extractos y fracciones obtenidas de cálices de *Physalis peruviana* L.', Biomédica, 27, 110 – 115.

Gallo F. (1992), 'Postharvest handling, storage and transportation of Colombian fruit', *Acta Hort,* 310, 155 – 169.

Galvis J. A., Fischer G. and Gordillo O. (2005), 'Cosecha y poscosecha de la uchuva', in n Fischer G., Miranda D., Piedrahíta W. and Romero J., Avances en cultivo, poscosecha y exportación de la uchuva (*Physalis peruviana* L.) en Colombia, Bogotá, Unibiblos, Universidad Nacional de Colombia, 165 – 190.

García H. R. Peña A. C. and García C. (2008), Manual de práctica de cosecha y acondicionamiento de la uchuva con fines de exportación, Bogotá, Corpoica.

Garg R. C. and Singh S. K. (1976), 'Effect of ethrel and cycocel on growth, flowering, fruiting behaviour and yield of cape gooseberry (*Physalis peruviana* L.)', *Progressive Hort,* 8 (3), 45 – 50.

Garzón C. P. and Villareal D. M. (2009), 'Efecto de algunas deficiencias nutricionales en la calidad poscosecha de la uchuva (*Physalis peruviana* L.)', Thesis, Bogotá, Facultad de Agronomía Universidad Nacional de Colombia.

Gordillo O. P., Fischer G. and Guerrero R. (2004), 'Efecto del riego y de la fertilización sobre la incidencia del rajado en frutos de uchuva (*Physalis peruviana* L.) en la zona de Silvania (Cundinamarca)', Agron Colomb, 22 (1), 64 – 73.

Guerrero B., Velandia M., Fischer G. and Montenegro H. (2007), 'Los ácidos carboxílicos de extractos vegetales y la humedad del suelo influyen en la producción y el rajado del fruto de uchuva (*Physalis peruviana* L.)', Rev Colomb Cienc Hortic, 1 (1), 9 – 19.

Gupta S. K. and Roy S. K. (1981), 'The floral biology of cape gooseberry (*Physalis peruviana* Linn., Solanaceae, India)', Ind J Agric Sci, 51 (5), 353 – 355.

Gustavo D Trinchero., Gabriel O Sozzi., Ana M Cerri., Fernando Vilella., and Adela A Fraschina. (1999), Ripening-related changes in ethylene production, respiration rate and cell-wall enzyme activity in golden berry (*Physalis peruviana* L.), a solanaceous species. *Postharvest Biology and Technology*, Volume 16, Issue 2, June 1999, Pages 139-145.

Gutierrez M. S., Trinchero G. D., Cerri A. M., Vilella F. and Sozzi, G. O. (2008), 'Different responses of goldenberry fruit treated at four maturity stages with the ethylene antagonist 1-methylcyclopropene', *Postharv Biol Technol,* 48, 199 – 205.

Hermann K. (1994), 'Über die Inhaltsstoffe und die Verwendung wichtiger exotischer Obstarten. VI. Solanaceen-Früchte'. Ind Obst-Gemüseverw, 6, 202 – 206.

Herrera A. (2000), 'Manejo poscosecha', in Flórez V. J., Fischer G. and Sora A. D., Producción, poscosecha y exportación de la uchuva (*Physalis peruviana* L.), Bogotá, Unibiblos, Universidad Nacional de Colombia, 109 – 127.

Icontec (1999), Frutas frescas. Uchuva. Especificaciones, Norma Técnica Colombiana NTC 4580', Bogotá, Instituto Colombiano de Normas Técnicas.

Icontec (2004), Frutas frescas. Uchuva. Especificaciones del empaque, Norma Técnica Colombiana NTC 5166', Bogotá, Instituto Colombiano de Normas Técnicas.

Kang, M.S. (1998). Using genotype-by-environment interaction for crop cultivar development. *Advances in Agronomy* 62:199-252.

Kaufmann M. R. (1972), 'Water deficits and reproductive growth', in Kozlowski T. T., Water Deficits and Plant Growth, Vol. 3, New York, Academic Press, 91 – 124.

Kays S. (2004), Postharvest Biology, Athens, Georgia, Exon Press.

Kirtikar, K. R. and Basu, B. D. (1935), Indian Medicinal Plants, Second edition (Published by Lalit Mohan Basu, Allahabad, India) Vol. II, p. 1492.

Klinac D. J. (1986), 'Cape gooseberry (*Physalis peruviana*) production systems', N Zealand J Exp Agric, 14, 425 – 430.

Klinac, D.J. (1986). Cape gooseberry (*Physalis peruviana*) production systems. *New Zealand Journal of Experimental Agriculture* 14:425-43.

Lagos T. C., Vallejo F. A., Criollo H. and Muñoz J. E. (2008), 'Biología reproductiva de la uchuva', *Acta Agron,* 57 (2), 81 – 87.

Lanchero O., Velandia G., Fischer G., Varela N. C. and García H. (2007), 'Comportamiento de la uchuva (*Physalis perunviana* L.) en poscosecha bajo condiciones de atmósfera modificada activa', Revista Corpoica Ciencia y Tecnología Agropecuaria, 8 (1), 61 – 68.

Legge A. P. (1974), 'Notes on the history, cultivation and uses of *Physalis peruviana* L.', *J Royal Hort Soc,* 99 (7), 310 – 314.

Lerner, I.M. (1954). *Genetic Homeostasis*. Edinburgh: Oliver and Boyd, UK.

Lieberei R. and Reisdorff C. (2007), 'Nutzplanzenkunde', Stuttgart, Georg Thieme, 179 – 180.

Ligarreto G. A., Lobo M. and Correa A. (2005), 'Recursos genéticos del género Physalis en Colombia', in Fischer G, Miranda D, Piedrahíta W and Romero J, Avances en cultivo, poscosecha y exportación de la uchuva (*Physalis peruviana* L.) en Colombia, Bogotá, Unibiblos, Universidad Nacional de Colombia, 9 – 27.

Lizana A. and Espina S. (1991), 'Efecto de la temperatura de almacenaje sobre el comportamiento en postcosecha de frutos de fisalis *(Physalis peruviana* L.)', *Proc Interamer Soc Trop Hort,* 35, 278 – 284.

López E. and Páez L. H. (2002), 'Comportamiento fisiológico de la uchuva (*Physalis peruviana* L.) bajo condiciones de refrigeración y películas plásticas para su conservación en poscosecha', Thesis, Bogotá, Facultad de Agronomía, Universidad Nacional de Colombia.

Leiva-Brondo, M., Prohens, J. and Nuez, F. 2011. Genetic analyses indicate superioty of performance of cape gooseberry (*Physalis peruviana* L.) hybrids. *J. New Seeds* 3(3):71-84.

Majumdar K. and Mazumdar B. C. (2002), 'Changes of pectic substances in developing fruits of cape-gooseberry (*Physalis peruviana* L.) in relation to enzyme activity and evolution of ethylene', *Scientia Hort,* 96, 91 – 101.

Marquez, C, Carlos J., Trillos G. Offelia., Cartagena V., Jose R., Cotes T and Jose M. (2009), Physico-chemical and Sensory Evaluation of Cape gooseberry Fruits. *Vitae [online].* Vol. 16, n.1, pp. 42-48.

Martínez F. E., Sarmiento J., Fischer G. and Jiménez F. (2008), 'Efecto de la deficiencia de N, P, K, Ca, Mg y B en componentes de producción y calidad de la uchuva (*Physalis peruviana* L.)', *Agron Colomb,* 26 (3), 389 – 398.

Mazorra M. F., Quintana A. P., Miranda D., Fischer G. and Chaparro M. de Valencia (2006), 'Aspectos anatómicos de la formación y crecimiento del fruto de la uchuva *Physalis peruviana* (Solanaceae)', *Acta Biol Colomb,* 11 (1), 69 – 81.

Mazorra M. F., Quintana A. P., Miranda D., Fischer G. and Chaves B. (2003), 'Análisis nsobre el desarrollo y la madurez fisiológica del fruto de la uchuva (*Physalis peruviana* L.) en la zona de Sumapaz (Cundinamarca)', *Agron Colomb,* 21 (3), 175 – 189.

Mazumdar B. C. (1979) 'Cape-gooseberry – the yam fruit of India', World Crops, 31 (1), 19 – 23.

McCain R. (1993), 'Goldenberry, passionfruit and white sapote: Potencial for cool subtropical areas', in Janick J., Simon J. E., New Crops, New York, Wiley, 479 – 486.

McCain, R. 1993. Golden berry, passion fruit, and white sapote: Potential fruits for cool subtropical areas. p. 479-486. In: J. Janick and J.E. Simon (eds.), New crops. Wiley, New York.

Menzel, Y.M. 1951. The cytotaxonomy and genetics of *Physalis.* Proc. *Am. Phil. Soc,*.95(2): 132-183.

Mercantila Publishers (1989), Guide to Food Transport: Fruit and Vegetables, Copenhagen, Mercantila Publishers.

Miranda Diego., Christian Utrichs and Gerhard Fischer, (2010), Imbibition and percentage of germination of Cape gooseberry (*Physalis peruviana* L.) seeds under NaCl stress. *Agronomia Colombiana* 28(1), 29-35.

Morton J. F. (1987), Fruits of Warm Climates, Miami, Julia F. Morton.

Morton, J. 1987. Cape gooseberry. p. 430–434. In: Fruits of warm climates. Julia F. Morton, Miami, FL.

National Research Council (1989), Lost Crops of the Incas, Washington, D.C., National Academy Press.

Novoa R., Bojacá M. and Fischer G. (2002), 'Determinación de pérdida de humedad en el fruto de la uchuva (*Physalis peruviana* L.) según el tipo de secado en tres índices de nmadurez', in Memorias IV Seminario de Frutales de Clima Frío Moderado, CDTF, nCorpoica, 20 – 22 November, Medellínm, 298 – 302.

Novoa R. H., Bojacá M., Gálvis, J. A. and Fischer G. (2006), 'La madurez del fruto y el nsecado del cáliz influyen en el comportamiento poscosecha de la uchuva (*Physalis peruviana* L.) almacenada a 12°C', Agron Colomb, 24 (1), 68 – 76.

Novoa, Rafael H., Bojaca, Mauricio., Galvis, Jesus Antonio and Fischer, Gerhard. (2006), Fruit maturity and calyx drying influence post-harvest behavior of Cape gooseberry (*Physalis peruviana* L.) stored at 12 °C. *Agron. Colomb.*[online]. Vol.24, n.1, pp. 77-86.

Nuez, F., J.J. Ruiz, and J. Prohens. (1997), *Mejora Genética para Mantener la Diversidad en los Cultivos Agrícolas*. Background study paper no 6. Rome: FAO.

Opara L. U., Studam C. J. and Banks N. H. (1997), 'Fruit skin splitting and cracking', *Hort Rev,* 19, 217 – 262.

Osterloh A., Ebert G., Held W-H., Schulz H. and Urban E. (1996), Lagerung von Obst und Südfrüchten, Stuttgart, Verlag Eugen Ulmer.

Pandey, K.K. (1957), Genetics of self-incompatibility in *Physalis ixocarpa* Brot.—A new system. *Am. J. Bot.,* 44: 879-887.

Perón J. Y., Demaure E. and Hannetel C. (1989), 'Les possibilities d'introduction et de developpement de solanacees et de cucurbitacees d'origine tropicale en France', *Acta Hort,* 242, 179 – 186.

Prohens, J. and F. Nuez. (1994). Aspectos productivos de la introduccion de nuevos cultivares de alquequenje (*Physalis peruviana* L.) en Espana. *Actas de Horticultura* 12:228-233.

Perry, L. M. and Metzger J. (1980), Medicinal Plants of East and Southeast Asia, nCambridge, MIT Press.

Quiros, C.F. 1984. Overview of the genetics and breeding of husk-tomato. *Hort. Science* 19(6):872-874.

Rao V. G. and Subramoniam V. (1976), 'A new post-harvest disease of cape-gooseberry', *J Univ Bombay,* 72, 58 – 61.

Rehm S. and Espig G. (1991), The Cultivated Plants of the Tropics and Subtropics, nWeihersheim, Germany, Verlag Margraf.

Restrepo, A. M., Cortés, M. and Suárez M. (2008), 'Evaluación sensorial de fresa (Fragaria x ananassa Duch.) y uchuva (*Physalis peruviana* L.) fortificadas con vitamina E', Rev Fac Nal Agr Medellín, 61 (2), 4,667 – 4,675.

Restrepo, Ana Maria., Cortes R., Miseal and Marquez, Carious Julio. (2009), Cape gooseberry (*Physalis peruviana* L.) minimally processed fortified with vitamin E. *Vitae [online],* vol. 16, n.1, pp. 19-30.

Rodríguez M. (2003), 'Estudio de la conservación de la uchuva (*Physalis peruviana* L.) utilizando los métodos de atmósfera modificada, refrigeración y encerado', Thesis, Bogotá, Departamento de Química, Facultad de Ciencias, Universidad Nacional de Colombia.

Rodríguez N. C. (2004), 'Estudio citogenetico en *Physalis peruviana* L.: 'uchuva' (Solanaceae)', Thesis, Bogotá, Departamento de Biología, Facultad de Ciencias, Universidad Nacional de Colombia.

Rodríguez S. and Rodríguez E. (2007), 'Efecto de la ingesta de *Physalis peruviana* (aguaymanto) sobre la glicemia postprandial en adultos jóvenes', *Rev Med Vallejiana* 4 (1), 43 – 53.

Sharma N. and Khan A. M. (1978), 'Fruit rots of cape gooseberry', *Indian Phytopathol,* 31, nb513 – 514.

Shiomi S., Kubo Y., Wamocho L. S., Koaze H., Nakamura R. and Inaba A. (1996), n'Postharvest ripening and ethylene biosynthesis in purple passion fruit', *Postharvest Biol Technol,* 8, 199 – 207.

Singh, D. B., S. Lal, N. Ahmad, S.N. Qureshi and A.A. Pal. 2011. Screening of capegooseberry (*Physalis peruviana*) collections for adaptation under temperate ecosystem. *Progressive Horticulture.* 43(2) 211-214.

Singh, D.B., A.A. Pal., Shiv Lal., NAzeer Ahmed and Anis Mirza. (2012), Growth and development changes of cape gooseberry fruits. *The Asian Journal of Horticulture.* 7 (2) 374-378.

Smith, Albert C. 1991. Flora Vitiensis nova: a new flora of Fiji. National Tropical Botanical Garden, Lawai, Kauai, Hawaii. Volume 5. 626 pp.

Steinmetz K. A. and Potter J. D. (1996), 'Vegetables, fruit, and cancer prevention: A review', *J Amer Diet Assoc,* 96 (10), 1,027 – 1,039.

Sykes, W. R. 1970. Contributions to the flora of Niue. New Zealand Department of Scientific and Industrial Research Bulletin 200. 321 pp.

Thompson J. F. (2002), 'Transportation', in Kader A. A., Postharvest Technology of Horticultural Crops, Publication 311, Oakland, University of California, Agriculture and Natural Resources, 259 – 269.

Torres, G., Neuhäuser, R., and Latham, D. W. 2001, in ASP Conf. Ser. 244, Young Stars Near Earth: Progress and Prospects, ed. R. Jayawardhana and T. P. Greene (San Francisco: ASP), 283.

Trillos O., Cotes J. M., Medina C. I., Lobo M. and Navas A. A. (2008), 'Caracterizacion morfológica de cuarenta y seis accesiones de uchuva (*Physalis peruviana* L.), en Antioquia (Colombia)', Rev Bras Frutic, 30 (3), 708 – 715.

Trinchero G., Sois G. O., Cerri A M., Vilella F. and Franschina A. (1999), 'Ripening-related changes in ethylene production, respiration rate and cell-wall enzyme activity in Goldenberry (*Physalis peruviana* L.) a solanaceous species', Postharvest Biol Technol, 16, 139– 145.

Ulloa, L.N., N.A. Vargas, D. Miranda and G. Fisher. (2006) Efecto de la salinidad sobre los parameters de desarrollo en species horticolas cultivades en sistemas sin suelo. pp. 53-76. In Florez, V.J., A, de la C. Fernandez, D. Miranda, B. Chaves and J.M. Guzman (eds.). Avances sobre fertirriego en la floricultura colombiana. Unibiblos; Facultad de Agronomia, Universidad Nacional de Colombia, Bogota.

USDA (2004), Treatment Manual, Food Safety and Inspection Service, United States Department of Agriculture.

Valencia M. L. (1985), 'Anatomía del fruto de la uchuva', *Acta Biol Colomb*, 1 (2), 63 – 89.

Verhoeven G. (1991), ' *Physalis peruviana* L.' in Verheij E. W. M. and Coronel R. E., Plant Resources of South-East Asia, Wageningen, Pudoc, 254 – 256.

Vietmeyer, N. (1991(. Lost crops of the Incas. *New Zealand Geographic* 10:49-67.

Villamizar F., Ramírez A. and Menes M. (1993), 'Estudio de la caracterización física, morfológica y fisiológica poscosecha de la uchuva (*Physalis peruviana* L.),' Agro-Desarrollo, 4 (1–2), 305 – 320.

Wagner, Warren L./Herbst, Derral R./Sohmer, S. H. 1999. Manual of the flowering plants of Hawaii. Revised edition. Bernice P. Bishop Museum special publication. University of Hawai'i Press/Bishop Museum Press, Honolulu. 1919 pp. (two volumes).

Wayama D.O., L.S. Wamocha., K.Ngamau and R.N. Ssonkko. (2006), Effect of Gibberrellic Acid on Growth and fruit Yield of Greenhouse-Grown Cape gooseberry. *African Crop Science Journal*, Vol. 14, no. 4, pp. 319-323.

Welsh, S. L. 1998. Flora Societensis: A summary revision of the flowering plants of the Society Islands. E.P.S. Inc., Orem, Utah. 420 pp.

Whistler, W. A. 1988. Checklist of the weed flora of western Polynesia. Technical Paper No. 194, South Pacific Commission, Noumea, New Caledonia. 69 pp.

Whitson M., Manos P. S. (2005), 'Untangling Physalis (Solanaceae) from the Physaloids: a two-gene phylogeny of the Physalinae', Syst Bot, 30, 216 – 230.

Wolff X. Y. (1991), 'Species, cultivar, and soil amendments influence fruit production of two *Physalis* species', *Hort Science*, 26 (12), 1558 – 1559.

Wonneberger C. (1985), 'Andenbeere – eine alte und neue Kulturpflanze', Gartenpraxis, 3, 60 – 61.

Wu S. J., Ng, L. T., Huang Y. M., Lin D. L., Wang S. S., Huang S. N. and Lin C. C. (2005), 'Antioxidant of *Physalis peruviana*', *Biol Pharm Bull*, 28, 963 – 966.

Wu S. J., Ng L. T., Lin D. L., Wang S. S. and Lin C. C. (2004), ' *Physalis peruviana* extract induces apoptosis in human Hep G2 cells through CD95/CD95L system and mitochondrial signalling transduction pathway, *Cancer Lett*, 215, 199 – 208.

Wu S. J., Tsai J. Y., Chang S. P., Lin D. L., Wang S. S., Huang S. N. and Ng L. T. (2006), 'Supercritical carbon dioxide extract exhibits enhances antioxidant and anti-inflamatory activities of *Physalis peruviana*, *J Ethnopharmacol*, 108, 407 – 413.

Yamaguchi M. (1983), World Vegetables. Principles, Production and Nutritive Values, Westport, AVI Publ.

Zapata J. L., Saldarriaga A., Londoño M. and Díaz C. (2002), 'Manejo del cultivo de la uchuv en Colombia', Rionegro, Antioquia, Corpoica C.I. La Selva. Boletín Técnico, 14, 1– 40.

Zapata J. L., Saldarriaga A., Londoño M. and Diaz C. (2005), 'Las enfermedades limitantes en cultivo y poscosecha de uchuva y su control', in Fischer G., Miranda D., Piedrahíta W. and Romero J., Avances en cultivo, poscosecha y exportación de la uchuva (*Physalis peruviana* L.) en Colombia, Bogotá, Unibiblos, Universidad Nacional de Colombia, 97– 110.

Index